AF885290

A Math Teacher's Workbook-style Study Guide to Help You Study for and Pass the Praxis II Mathematics Content Examination with Practice Problems and Detailed Testing Strategies

Second Edition

Written by Kyle J. Kirby

Published by the Easy As π Review Series

Praise from the first edition:

"The examples and explanations are great in this book. I saw at least two (if not more) similar problems on my exam. The author also gives tips on how to solve different problems via a graphing calculator. Best of all, I PASSED--on the first try! :)
For those who do not specialize in math, or have not completed recent coursework: It does not go as in depth to all the subjects of the test as other books, but it was sufficient for me. I am a mathematics & mathematics education double major, and I took the test a year after I completed all of my coursework."

TeachMath12, 5 star review on Amazon.com

"As someone looking just for practice problems for the Praxis II math exam, this book was exactly what I was looking for. I took the exam today and the problems and topic areas were on target with the test. I would recommend this for anyone looking for a quick refresher and plenty of example problems."

Kacie, 5 star review on Amazon.com

"I had taken the Praxis 2 several times and had not passed. I got this book the week before I was scheduled to take the test and I finally passed! I would recommend this book to anyone."

Hlaughorn, 5 star review on Amazon.com

"I just went through 0061 and I feel this one is the most effective review guide for the test. It is more close to the real test than any other book I used. Some other books on this topic just state a lot what will be tested in the test that is useless in taking the test which is solving problem all the way. In this book, the author gave a lot examples and some practice that directly like what is in the test. If you do not have much time, do not waste time searching like I did, just pick this one … There is little choice. No other review guide for 0061 is better than this one."

Wei Qui, 5 star review on Amazon.com

"This book is a gem. I am taking the test in about a month, and this book allowed me to jump right into problem solving and practice. It quickly moves across the range of topics, which gave me the opportunity to assess my weaknesses and strengths. The discussion of each solution is excellent, suggesting more than one approach in many examples. And several of the examples are pretty humorous and fun!"

Mark W. Frazier, 5 star review on Amazon.com

"The book is a great resource for test taking tips - but I would use it in conjunction with one of the others for the practice tests. There is only one VERY short practice quiz in this book, but there are lots of examples."

praxistaker, 3 star review on Amazon.com

Pipe Dream Industries, L.L.C
email: kyle.j.kirby@gmail.com

Copyright © 2013 Pipe Dream Industries, publishing as the Easy As π Review Series.

All rights reserved. This publication is protected by Copyright and permission should be obtained from Pipe Dream Industries, L.L.C. prior to any commercial use or redistribution.

ISBN 10-digit: 0983902631

ISNB 13-digit: 978-0-9839026-3-8

Second Edition

Table of Contents

Chapter 1

Introduction

Dear Reader,

When I studied for the Praxis II ® (0061) exam for mathematics content knowledge, I couldn't find any effective study guides and I could hardly find any challenging practice problems; that is why I wrote this guide.

In writing this study guide book, I strove to create difficult problems that you may find on the actual exam. I included detailed answers to every problem; hopefully through reading about different topics and then attempting related practice problems, you can improve your chances of passing the Praxis II ® (0061) exam for mathematics content knowledge.

This is a second edition book. Over the past couple of years I have collected feedback from over a dozen readers, changed many things, fixed typos, followed advice to add content here or there and I am pleased with the result. I hope you are as well. If you find this book useful, please review it on Amazon. If you have suggestions or any questions at all as many have, please email me at kyle.j.kirby@gmail.com. I usually respond within hours, if not days.

Good luck on the exam,

Kyle J. Kirby

1.1. About the Author

I believe that the credentials of the author are important if you are to trust in his or her advice about test taking. I have taught all subjects of high school mathematics in Colorado: Algebra I, Geometry, Algebra II, Precalculus, Calculus, Trigonometry, and Statistics. I passed the Praxis II ® (0061) exam for mathematics content knowledge as well. I am not a perfect mathematician, but I am a good test taker and while writing this guide, I strove to teach concepts well through the use of straightforward explanations and through the use of examples.

1.2. About the Praxis II ® Mathematics (0061) Content Knowledge Exam

The Praxis II Mathematics Content exam is 50 questions long and the testers allow only 2 hours to take the exam:

$$2\,hours\,x\frac{60\,minutes}{hour}x\frac{1}{50\,questions}=2.4\frac{minutes}{question}$$

The Praxis II ® (0061) is a difficult exam; to say otherwise would be misleading. You will almost certainly be pressed for time which is why mastering the techniques that I discuss in the testing strategies section of this book is essential. Every second counts.

Here is the breakdown of percentages based on topic:

Content	**Approximate Percentage on Exam**
Algebra and Number Theory	16%
Measurement	6%
Geometry	10%
Trigonometry	8%
Functions	16%
Calculus	12%
Data Analysis and Statistics	10-12%
Probability	4-6%
Matrix Algebra	8-10%
Discrete Mathematics	6-8%

The information in the table above comes directly from the Praxis II website.

Chapter 2

Frequently Asked Questions

Question:

Is the Praxis II ® (0061) Mathematics Content Knowledge Exam the exam that I need to take?

Answer:

If you want to be a secondary mathematics teacher in the United States, you probably need to pass the Praxis II Mathematics (0061) Content Knowledge Exam. There may be some states that do not require you do pass the exam; please refer to the department of education standards to find out if your state requires you to pass the exam in order to obtain licensure.

Question:

How is the test given?

Answer:

The exam is given in both on paper or on a computer. If you take the computer version, they provide an on-screen calculator, or you can use your own.

Question:

Can I use a graphing calculator on the exam?

Answer:

Yes, you can. You may use any graphing calculator on the exam (and are expected to bring one). You may not use a graphing calculator with a typewriter keyboard (which includes the TI-92 Plus and the Voyage 200). You may not bring any sort of computer to the exam.

I only discuss calculator techniques for the TI-83 – which works the same as the TI-82's, the other TI-83's, the TI-84's, and the TI-85 models. If you need help learning how to use other models, you will have to look beyond my guide.

Question:

What should I bring to the exam?

Answer:

Bring several sharpened #2 pencils, your graphing calculator, your registration information, photo identification, and snacks to eat before the exam.

Question:

Where do I register for the Praxis II ® (0061) exam?

Answer:

You can register for the exam at www.ets.org.

Question:

What should I eat before the exam?

Answer:

Eat a big breakfast with lots of protein (which provides longer lasting energy).

Question:

What if I don't pass?

Answer:

You may take the exam as many times as you would like without penalty (except that you have to pay to take the exam again and, of course, you have to study for it again).

Chapter 3

Testing Strategies

3.1. Estimate Before You Answer

When I approach a math problem first through **estimation**, often times I can catch my own mistakes. Here is an example. Try to estimate your answer first.

Example 3.1.a: Which equation is **normal** to the function $f(x) = x^2$ at $x=3$?

A. $y = 6x - 5$

B. $y = 6x - 6$

C. $y = -\frac{1}{6}x + \frac{19}{2}$

D. $y = -\frac{1}{6}x + 10$

Solutions to Example Problems:

Solution to 3.1.a:

The answer is C. This is a complicated problem that involves using calculus to find the slope of the line **tangent** to the function $f(x)=x^2$ and then using basic algebra to find the line **perpendicular** to that function at x = 3 (note that "normal" means perpendicular).

For our estimation, we can look at the curve and estimate the slope of the normal line right off the bat. Here is the function on the left and the function with a normal line drawn in on the right:

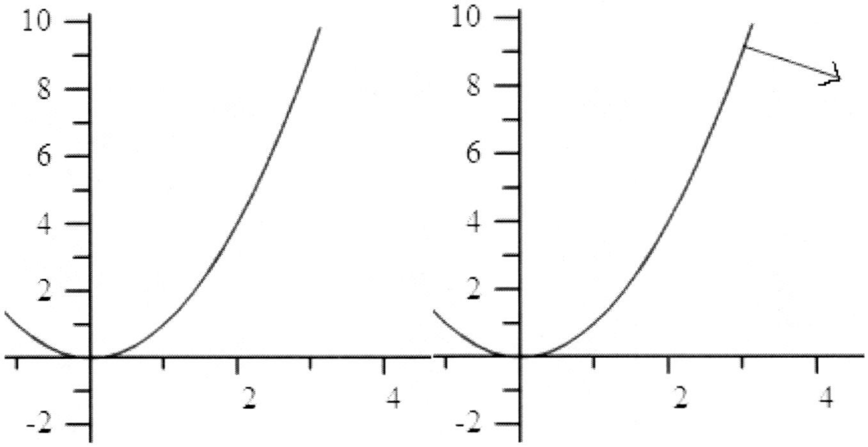

Notice that I drew the normal line starting from the point on the function where $x=3 or f(3)$ and I made the line perpendicular to the function. Based on that line above, I would estimate that my answer is $y=-\frac{1}{4}x+11$ because the slope looks roughly -1/4 and the y-intercept looks like it is at about 11. Now let's move on to the second testing strategy: **Process of Elimination** before we answer the question above.

3.2. Use Process Of Elimination (POE)

Since we estimated the normal line, we can cross out answers A and B because they each have positive slopes. Thus, even if you don't remember how to find the normal line using the first **derivative** plugging in a point, you still have a 50% chance of getting this question right.

On the exam, using process of elimination and estimation to limit your answer choices is key to maximizing your score. Do everything you can do cross off answers that don't make sense.

The actual answer to the example problem must be solved using calculus. We start by finding the **slope** of $f(x)=x^2$ at the point $f(3)$; we do so by finding the first derivative of $f(x)=x^2$ which is $f'(x)=2x$ and then we plug in the point $x=3$. Thus, $f'(3)=2(3)=6$. Therefore, the slope of the line $f(x)=x^2$ at x=3 is 6 and the slope of the normal line (the perpendicular line) must be -1/6.

So far this is what we know about the normal line: $y=-\frac{1}{6}x+b$. To find b (the y-intercept), we need to plug in a known point. We know one point on that line because it shares a point with the function $f(x)=x^2$; that point is (3,9). I found this point by plugging 3 into the function $f(x)=x^2$ (which gives us 9 as a y-coordinate)

Next, we plug in 3 for x and 9 for y into $y=-\frac{1}{6}x+b$:

$$9=-\frac{1}{6}(3)+b \Rightarrow 9=-\frac{1}{2}+b \Rightarrow b=9\frac{1}{2} \Rightarrow b=\frac{19}{2}$$

Therefore, the line normal to the function $f(x)=x^2$ at x=3 is $y=-\frac{1}{6}x+\frac{19}{2}$.

3.3. Have Resilience

By resilience, I mean the ability to bounce back from questions that you know that you got wrong or had to guess at. The Praxis II ® (0061) exam is a difficult exam and it is 50 questions long which means that you need all your mental power to finish. When I reached the end of the exam, I was exhausted (and hungry for ice cream!).

Of the hundreds of math students that I have taught over the years, the most successful testers have been the resilient ones: the ones that did not let a difficult problem phase them during an exam. If you get to a problem that you can't solve: estimate, use process of elimination or work backwards if possible, and guess if you have to, but move on after each problem. The Praxis II ® (0061) exam WILL stump you; remember, you don't need to get every question correct to pass, but if you give up half way through the exam because you feel like you can't pass, then you will not pass.

Be resilient like a warrior. When you get knocked down, get back up. I remember this scene in the movie *The Last Samurai* where the character played by Tom Cruise is getting a serious beat down; Tom challenges this samurai to duel after duel and gets knocked into the mud over and over again. Little samurai children are watching Tom; they want him to give up. Every time Tom gets knocked down, he gets back up. You must be this character to pass the Praxis II ® (0061) exam. You must not listen to the samurai children when they tell you to stay in the mud; get up after a difficult question and start the next question with a clear head.

3.4. Work Backwards (Plug In Answers)

In some questions, the test makers leave the option open for you to either plug in answer choices or to make up numbers and plug them into the answer choices; such techniques can make complicated problems simple. In general, when the exam asks you to create an algebraic equation, you can probably **work backwards**. Also, if you encounter a difficult percentage problem, make up easy numbers to plug in (like 100). To find each question where I use this technique, look up "work backwards" in the index. The best way for me to explain this is to show another example:

Example 3.4.a: Luke and Doni can each mow 7 lawns per day. Luke charges $20 per mow while Doni charges $15 per mow. If Doni mows lawns for 50 straight days, how many days will Luke have to mow lawns in order to make more money than Doni?

A. 36 days
B. 37 days
C. 38 days
D. 39 days

Solutions to Example Problems:

Solution to 3.4.a:

The answer is C. If you don't know how to solve this problem using algebra, plug in the answers until you get a value higher than what Doni made in 50 days. First, Doni makes $\frac{7 lawns}{day} x \frac{\$15}{lawn} x 50 days = \$5250$ (notice how I used stoichiometry to keep track of which units canceled).

This means that Luke will need to make more than that. Here is a trick: start with one of the middle numbers – that way if you find that Luke needs to mow more days than the number you picked, you can cancel out all the answers that would give him less money. Let me show you what I mean:

I Start with 37 days: if Luke mows for 37 days, he makes $37 days\, x\, 7\frac{lawns}{day} x \frac{\$20}{lawn} = \$5180$. Since Doni makes more money than that, we need a higher number which means we can cancel out both answer B: 37 days AND answer A: 36 days. We just gave ourselves 15 seconds by starting at 37 rather than at 36.

Let's try 38 days: if Luke mows for 38 days, he makes $38 days\, x\, 7\frac{lawns}{day} x \frac{\$20}{lawn} = \$5320$. There is our correct answer.

Here is how to solve the problem using algebra: Luke must mow for:

$$7\frac{lawns}{day} x \frac{\$20}{lawn} x D days > \$5250 or \frac{\$140}{day} x D > \$5250 or D > \frac{\$5250 x days}{\$140}$$

$$or D > 37.5 days$$

Note that I used an inequality to represent that Luke needed to make MORE money than Doni. Luke is so competitive!

Chapter 4

Foundations of Mathematics

A quick word on this chapter – many of the concepts in this chapter are "easier" than in other chapters, but between the algebra topics and matrices, this chapter covers 22-24% of the test. Since a very difficult calculus question weighs the same as a question on basic algebra, you should take the points wherever you can.

Passing this exam isn't about getting the hardest questions correct (although that helps), it's about racking up points wherever you can.

I included many questions that you might think are too simple and you may be tempted to skip them. My advice is to get really good at the basic stuff so that when you come across an algebra question on the exam, you can recognize it, execute it perfectly and without time wasted, and spend your extra time figuring out the truly difficult questions with the time you saved.

4.1. Ordering Finite Sets and Recognizing Equivalent Forms of a Number

The most likely application of this topic on the exam is for ETS to ask you to take a set and order them from least to greatest or greatest to least or to place numbers on a number line. However, just to be safe, let's learn what "order" can also mean (even though I have literally never been asked this on any exam):

The order of a finite set of real numbers is the number of elements in a set. For example, the set $\{2,4,6,8,10\}$ has an order of 5 because there are 5 elements in the set.

Infinite sets have infinite order. An example of an infinite set is: $\{1,2,3,4,\ldots\}$ where the … means that the set goes on forever.

Now back to the more likely chance that you will have to order numbers from least to greatest or greatest to least: real numbers can have a variety of different forms. Here are a few:

- .053 can be written using scientific notation as $5.3x10^{-2}$.

- All rational numbers can be written as a quotient of two integers (but the denominator cannot be zero).
- All real numbers can be written as a number to the exponent 1. For example, $3 = 3^1$.
- Any number to the zero power is simply 1. For example, $15^0 = 1$.
- Fractions can be written in many forms which are equivalent whether they are reduced or not. For example, $\frac{3}{7} = \frac{6}{14} = \frac{9}{21} \ldots$
- To move between percent and decimal form, divide the percent by 100. For example, 4.5% is actually .045 in decimal form.

Example 4.1.a: Use Scientific Notation to simplify: $\dfrac{(1.24x10^{-6})(2.62x10^{10})}{(1.21x10^3)}$

A. $2.68x10^2$

B. 268

C. 26.8

D. 2.68

Example 4.1.b: Which is equivalent to .000000764?

A. $7.64x10^{-7}$

B. $7.64x10^7$

C. $10x7.64^{-7}$

D. $10x$ .000000764

Example 4.1.c: Order the following numbers from least to greatest: the positive square root of 5, $2.23, 2\frac{22}{100}, 2.23x10^{-1}, \frac{16}{7}$

A. $2.23x10^{-1}, 2\frac{22}{100}, 2.23, \sqrt{5}, \frac{16}{7}$

B. $\sqrt{5}, 2.23x10^{-1}, 2\frac{22}{100}, 2.23, \frac{16}{7}$

C. $\frac{16}{7}, \sqrt{5}, 2.23x10^{-1}, 2\frac{22}{100}, 2.23$

D. $2.23x10^{-1}, \sqrt{5}, 2\frac{22}{100}, 2.23, \frac{16}{7}$

Solutions to Example Problems:

Solution to 4.1.a:

The answer is C. This problem is most easily found by plugging the exact numbers into your calculator. On your graphing calculator, it will look like this:

$$(1.24_E\text{-}6)(2.62_E10)/(1.21_E3)$$

where the "$_E$" means "times ten to the" whatever power. For example, $(1.24_E\text{-}6)$ means 1.24 times 10 to the -6th power.

Solution to 4.1.b:

The answer is A. Every place a decimal point moves right gives us a 10^{-1} (dividing by 10) while every place that a decimal point moves to the right is a multiplication by 10.

Solution to 4.1.c:

The answer is A. $2.23x10^{-1}$ is equivalent to .223; $2\frac{22}{100}$ is equivalent to 2.22; 2.23 is itself; $\sqrt{5}$ is equivalent to 2.236; and $\frac{16}{7}$ is equivalent to 2.2857.

4.2. Ratios and Proportions

Use cross-multiplication to solve ratios. I will use an example to demonstrate how to set up proportions. Say that you have the statement: "Luke likes ponies as much as Doni likes candy." To set up a proportion, you put corresponding parts in corresponding locations as such:

$$\frac{Luke}{Ponies} = \frac{Doni}{Candy} \; or \; \frac{Luke}{Doni} = \frac{Ponies}{Candy} \; or \; \frac{Ponies}{Luke} = \frac{Candy}{Doni}$$

Hopefully you can see that it doesn't matter where your corresponding parts go so long as they match up equivalently with other corresponding parts.

Example 4.2.a: A television has a 4:3 ratio in width to height. Its total area is 1200 square inches. How wide would a television need to be to have the same total area, but with a 16:9 width to height ratio?

A. 144

B. $3\sqrt{80}$

C. $\sqrt{1200}$

D. $80/\sqrt{3}$

Example 4.2.b: What value of x makes the equation $\frac{5+x}{6}=\frac{2}{3}$ true?

A. $x=1$

B. $x=3$

C. $x=-1$

D. $x=1/4$

Example 4.2.c: Luke is making Sloppy Joes. He knows that the correct ratio of ketchup to mustard in the Sloppy Joe sauce is 6:1 ketchup: mustard. How much ketchup should he use to make 8 oz. of Sloppy Joe sauce?

A. $\frac{48}{7}$ oz.

B. 10 oz.

C. $\frac{66}{7}$ oz.

D. $\frac{71}{6}$ oz.

Solutions to Example Problems:

Solution to 4.2.a:

The answer is D. Let's check the answer choices by working backwards:

If the width is 144 (answer choice A), then to find the corresponding height, I set up a proportion: $\frac{width}{height} = \frac{144}{h} = \frac{16}{9}$. Solving for h, we get $144(9) = 16h$ or h = 81. Therefore the overall area would be $144x81 = 11664 \, square\, inches$; this is obviously not correct since the area must be 1200.

If the width is $3\sqrt{80}$ (answer choice B), then the height must be $3\sqrt{80}(9) = 16h$ or h = 15.093 inches. Therefore the area would be $15.093x3\sqrt{80} = 405 \, square\, inches$.

If the width is $\sqrt{1200}$ (answer choice C), then the height must be $\sqrt{1200}(9) = 16h$ or h = 19.486. Therefore the area would be $19.486x\sqrt{1200} = 674 \, square\, inches$.

The answer should be D, but let's check anyway. If the width is $\frac{80}{\sqrt{3}}$, then the height must be $80/\sqrt{3}(9) = 16h$ or h = 25.98 inches. Therefore the area would be $25.98x\frac{80}{\sqrt{3}} = 1200 \, square\, inches$.

Alternatively, you can set up an algebraic equation to represent the problem given. Basically, the product of two numbers is 1200. Those two numbers must have a 16:9 ratio between them. Thus, $wh = 1200 \, and \, \frac{w}{h} = \frac{16}{9}$. Solving the proportion for h we get $16h = 9w \Rightarrow \frac{9w}{16} = h$. Next we substitute $\frac{9w}{16}$ in for h:

$$(w)\left(\frac{9w}{16}\right) = 1200 \Rightarrow 9w^2 = 19200 \Rightarrow w^2 = \frac{19200}{9} \Rightarrow w^2 = \frac{6400}{3}. \text{ Thus, } w = \frac{\sqrt{6400}}{\sqrt{3}} \Rightarrow w = \frac{80}{\sqrt{3}}.$$

Solution to 4.2.b:

The answer is C. Take $\frac{5+x}{6} = \frac{2}{3}$ and cross multiply. $3(5+x) = 12 \Rightarrow (5+x) = 4 \Rightarrow x = -1$.

Solution to 4.2.c:

The answer is A. First you set up the following ratios just to make sure you have it straight:

$$\frac{ketchup}{mustard} = \frac{6}{1} \, and \, \frac{ketchup}{total\, sauce} = \frac{6}{7}$$

Next, substitute the total number of ounces you need into the total sauce part of the ratio:

$$\frac{ketchup}{8} = \frac{6}{7}$$

Cross multiply, and solve for ketchup:

$$7(ketchup) = 48 \Rightarrow ketchup = \frac{48}{7}.$$

4.3. Expressions with no Solutions or Infinitely Many Solutions

Occasionally, you may encounter algebraic equations that have no solutions or an infinite number of solutions. An algebraic equation will have no solutions if, after you simplify, you get a false statement (e.g. 3=5 is a false statement). If you simplify an algebraic equation and receive a true statement (e.g. 4=4 is a true statement), then the equation will have an infinite number of correct solutions.

Example 4.3.a: Which expression represents the following statement: the product of $\frac{2}{3}$ and the absolute value of the difference between 9 and -4?

A. $\dfrac{\frac{2}{3}}{|9-4|}$

B. $\dfrac{\frac{2}{3}}{|9-(-4)|}$

C. $\frac{2}{3}|9-4|$

D. $\frac{2}{3}|9-(-4)|$

Example 4.3.b: A graph with two parallel lines contains how many solutions?

A. 1 solution
B. Infinitely many solutions
C. No solutions
D. Not enough information given

Example 4.3.c: The equation $2\left(x+\frac{3}{2}\right)=\frac{4x+6}{2}$ contains how many solutions?

A. 1 solution
B. Infinitely many
C. No solutions
D. 2 solutions

Solutions to Example Problems:

Solution to 4.3.a:
The answer is D.

Solution to 4.3.b:
The answer is C. When you are given two lines on a graph, the "solution" is the place where the lines intersect; this is also the value of x where, when plugged in for x, will yield the same y-value for each equation. With parallel lines, the lines will never cross which means that there will never be a value of x that will be true for each of the two lines.

Solution to 4.3.c:
The answer is B. Let's simplify and see what happens:

$2\left(x+\frac{3}{2}\right)=\frac{4x+6}{2}$ Multiply both sides by 2.

$4\left(x+\frac{3}{2}\right)=4x+6$ Then distribute the 4 through the left hand side.

$4x+6=4x+6$ At this point you should already be able to see that every value of x will make the equation true. This means that the equation has an infinite number of solutions.

4.4. Creating Algebraic Expressions and Equations from Verbal Descriptions

Here is a quick rundown of how different words translate to different mathematical symbols:

Word	Symbol
Of, Product	Multiplication
Per, Divided by, Quotient	÷
Is, Equals, Results in	=
A number	Any variable, such as x
Quantity	Parentheses
Less, Minus, Subtracted from, Difference	–
Added to, The sum of, Greater	+
The opposite of	The negative of something

Example 4.4.a: The sum of five and the product of a number and six is three. What is six less than four times the number?

A. $-\frac{1}{3}$

B. $-\frac{22}{3}$

C. -10

D. $-\frac{18}{3}$

Example 4.4.b: Five times a number is fifty-six greater than the product of two and the opposite of the number. What is the number?

A. 5
B. 6
C. 7
D. 8

Solutions to Example Problems:

Solution to 4.4.a:

The answer is B. First we write the equation: $5+6n=3$. Solving for n, we get $n=-\frac{1}{3}$. Then we write the second equation noting that 6 is subtracted from the product of 4 and $-\frac{1}{3}$: $4\left(-\frac{1}{3}\right)-6$.

Then, we simplify: $-\frac{4}{3}-6=-\frac{4}{3}-\frac{18}{3}=-\frac{22}{3}$.

Solution to 4.4.b:

The answer is D. Here is my translation of words to symbols:

$$5n=56+2(-n)$$

$$5n=-2n+56$$

$$7n=56$$

And so $n=8$.

4.5. Numerical Problem-Solving Strategies

Many algebraic concepts can be represented geometrically. Algebra tiles, for example, help students make geometric manifestations of algebraic concepts.

Consider x^2, which is a common algebraic expression. Geometrically, x^2 represents a two dimensional area (a square with a side length of x).

The Praxis II ® (0061) exam will make you write algebraic formulas to solve problems, which can be complicated. However, many of these types of problems can be solved by plugging in actual numbers and then working backwards; only one of the solutions will fit with the numbers you chose.

The problem below is a geometric representation of an algebraic concept. I will show you how to solve the problem below by picking numbers to demonstrate this method.

Example 4.5.a: Which algebraic equation models the shaded area below?

A. cx + bx + ab
B. b(x + a) + cx
C. ba + x(c + b)
D. cd – a(c-b)

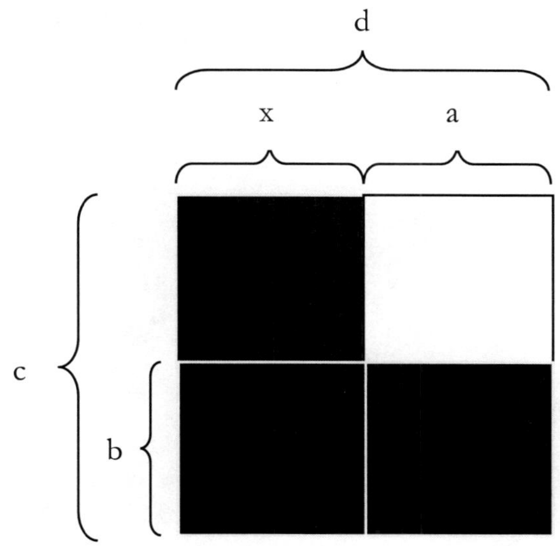

Solutions to Example Problems:

Solution to 4.5.a:

The answer is D. To solve this by picking values for my unknowns, I let b=1; x=1; a=1; c=2; and d=2. I know that the total area should be 4 based on the numbers I picked.

Therefore:
Answer A = 2+1+1 = 4, which is wrong.

Answer B = 1(1+1)+ 2 = 4, which is wrong.

Answer C = 1+ 1(2+1) = 4, which is wrong.

Answer D = 4 – 1(2 – 1) = 3, which is correct.

Alternatively, you can solve this problem using algebra and process of elimination:

If A is the answer, then cx (the left 2 blocks) and bx (the left bottom box) overlap. Answer A must be incorrect.

If B is the answer, then once again the bottom left box is counted twice.

If C is the answer, then an extra box is counted because x(b+c) is actually 3 boxes. Answer D represents the entire area (c times d) minus the top left box a(c-b). Therefore D is correct.

4.6. Using Technology to Problem-Solve

For many problems on the Praxis II ® (0061) exam, you can solve them more quickly if you have good command of a graphing calculator. I wrote the problem below to show you a specific example of an extremely difficult problem made simple with the use of a graphing calculator.

Example 4.6.a: Find the zeroes of the function $y(x)=x^4+2x^3-9x^2-2x+8$

A. $x=-4,1,2,-1$

B. $x=1,2,-9,-2,8$

C. $x=-1,-2,9,2,-8$

D. $x=3,2,-1,-4$

Solutions to Example Problems:

Solution to 4.6.a:

The answer is A. This equation is unfactorable by any methods you may have learned. By far, the easiest and quickest way to find the factors is to plug $y = x^4 + 2x^3 - 9x^2 - 2x + 8$ into your graphing calculator and look for all the places where the function crosses the x-axis.

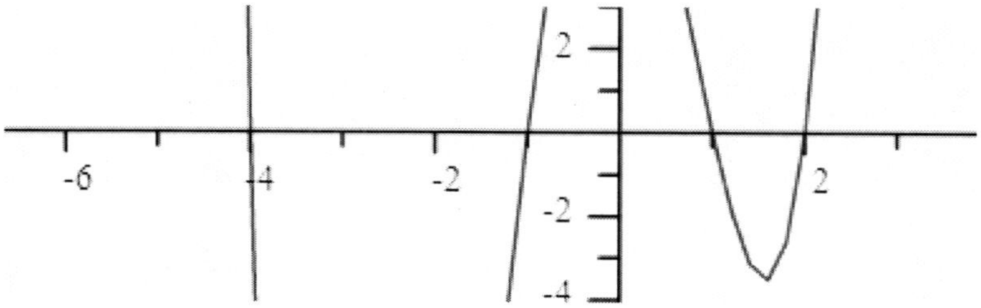

As you can see from the image above, the function crosses the x-axis at -4, -1, 1, and 2 which means that those numbers are the factors (zeroes) of the given function.

4.7. Subsets of Real Numbers

Here are some of the categories which the Praxis II ® (0061) exam creators want you to you know well: **integers**, **prime numbers**, rational and irrational numbers, real numbers, and imaginary numbers.

Here is a set of each:

integers: $\{1,2,3,4\}$

prime numbers: $\{1,2,3,5,7,11,13\}$

rational numbers: $\{1,\frac{1}{2},\frac{1}{4},\frac{1}{5},\frac{4}{3}\}$ (note, each can be expressed as a fraction of 2 numbers)

irrational numbers: $\{\pi, e, \sqrt{2}, 1.\overline{12}\}$ (each cannot be expressed as a quotient)

real numbers: $\{-3,\frac{2}{3},\sqrt{2},5\}$ (each represents a quantity along a continuum, includes rational and irrational numbers).

imaginary numbers: $\{\sqrt{-1},\sqrt{-7},\sqrt{-10}\}$

The examples below test your knowledge of **unions** and **intersections** as well as your knowledge of integers and real numbers.

Example 4.7.a: Which of the following represents an intersection between integers and real numbers?

A. {45,46,47}

B. {44.5,45,45.5}

C. Both A and B

D. Neither A nor B

Example 4.7.b: Which of the following represents a union between integers and real numbers?

A. {45, 46, 47}

B. {44.5, 45, 45.5}

C. Both A and B

D. Neither A nor B

Solutions to Example Problems:

Solution to 4.7.a:

The answer is A. An intersection means "and"; an integer is a whole number, and a real number is any number that is not imaginary (this constitutes rational, and irrational numbers). Since this question is looking for an intersection of two sets, the resulting set will contain all the numbers that are both integers AND real numbers. Since the set in answer B contains numbers which are not integers (44.5 and 45.5), answer B must be incorrect.

Solution to 4.7.b:

The answer is C. A union means "or". The numbers in the correct set(s) can either contain integers, real numbers, or a combination thereof. The set in answer A and the set in answer B each fit the requirements to be unions of real numbers and integers.

4.8. Considering Context when Using Numbers

Oftentimes, certain numbers just do not make sense in the context of a problem. For example, every number has a positive and a negative square root; if the Praxis II ® (0061) exam asks you to determine the possible values for the hypotenuse of a triangle, the answers will not be negative because a hypotenuse cannot have a negative value.

When using dollars and cents, it makes sense to round to the nearest cent (unless they're asking about British ha'pennies right?). Speaking of...

Example 4.8.a: The British halfpenny coin represented $1/480^{th}$ British Pound Sterling. If one British pound sterling is worth 1.5674 U.S. Dollars. How many halfpennies would 5 dollars be worth today?

A. 3761.8
B. 3762
C. 1531.2
D. 1531

Solutions to Example Problems:

Solution to 4.8.a:

The answer is D. I use stoichiometry and conversion factors to solve this problem to keep track of units. Note that every time you want to cancel a current unit, use a conversion factor with the unit that you want to cancel opposite its current place; for example, if dollars are currently on the top of a fraction, place dollars on the bottom of the next fraction to cross cancel dollars.

$$5\,dollars \times \frac{1\ Pound\ Sterling}{1.5674\,Dollars} \times \frac{480\,half\ pennies}{1\,British\ Sterling} = 1531.198\,halfpennies$$

Here is where it makes sense to round to the nearest halfpenny. The answer should be 1531 halfpennies.

4.9. Algebraic Operations

Here are the operational properties that you will need to know:

distributive property: $a(b+c) = ab + ac$

multiplication property of zero: $0a = 0$

multiplicative inverse property: $a\left(\frac{1}{a}\right) = 1$

additive inverse property: $a + (-a) = 0$

additive identity property: $a + 0 = 0 + a = a$

associative property:
$(a+b)+c = a+(c+b)\, or\, a(bc) = (ab)c(a+b)+c = a+(c+b)\, or\, a(bc) = (ab)c$

commutative property: $a+b = b+a\, or\, ab = ba$

More on using the properties in the following section.

Example 4.9.a: What does y equal in the following equation? $(x+4)^2 = x^2 + 8x + y + 16$

A. 0
B. 1
C. 2
D. 4

Solutions to Example Problems:

Solution to 4.9.a:

The answer is A. Expand the left and start subtracting parts out from each side:

$$(x+4)^2 = x^2+8x+y+16$$

$$x^2+8x+16 = x^2+8x+y+16$$

If y = 0, the left and right sides are equal to one another.

4.10. Closure, Commutative, Associative, and Distributive Properties with Numbers

---

The **closure property** is used to describe sets that are "closed" under a certain type of operation. To be closed under an operation, you must be able to take any two numbers from a set, use the operation on them to get another number in the set.

For example, say that our set is the set of all odd integers: $\{\pm1, \pm3, \pm5, \ldots\}$. To be closed under normal multiplication, we have to be able to take any two elements in the set, multiply them and get another element in the set. Let's try a couple:

$-1\times3=-3$ and -3 belongs to the set of odd integers
$5\times7=35$ and 35 belongs to the set of odd integers
...

To prove that the set of odd integers is closed under multiplication we would have to do a formal proof, but you should be able to see that any odd integer times any other odd integer will result in an odd integer. Therefore, the set of odd integers is "**closed**" under normal multiplication; in other words, it has the closure property.

Is the set of integers closed under normal addition? Remember, to be closed under an operation (like addition) any odd integer plus any other odd integer must result in an odd integer.

Counter example: $7+7=14$ and since 14 is not an odd integer, the set of odd integers is NOT closed under normal addition.

Commutativity is the property that allows us to change the order in an equation. I remember the commutative property because it makes me think of commuting to work; that is, traveling to work. A math equation is commutative if you can let part of it travel.

For example, $4x+7=7+4x$. The terms 4x and 7 are commutative because their order doesn't matter. In the normal algebra that we study in middle school and high school, almost everything is commutative. Can you think of anything that isn't commutative? The best example of an example of a non-commutative situation is the multiplication of matrices. However, multiplication of matrices is beyond the scope of the Praxis II ® (0061) exam.

The **associative** property determines whether or not you can re-associate parts of a math equation. When you determine associativity, ask the question: $does\,(ab)c = a(bc)$?

If a set has the **distributive property** then $a(b+c) = ab + ac$.

Example 4.10.a: Which of the following is a representation of the commutative property?

A. $a(b+c) = ab + ac$
B. $a(bc) = (ab)c$
C. $abc = acb$
D. $ac + ab = a(c+b)$

Solutions to Example Problems:

Solution to 4.10.a:

The answer is C. Answer C contains the only situation where variables commute to other places. Answer A represents the distributive property; answer B represents the associative property; answer D represents the distributive property.

4.11. Closure, Commutative, Associative, and Distributive Properties with Newly Defined Operations

---

I am going to discuss this topic through example. If you need to review the concepts of closure, commutativity, associativity, or distributivity, those concepts are discussed in section 4.3.

You may have not seen any problems regarding "newly defined operations" unless you recently took a class in abstract algebra. The concept is to use an operation BESIDES normal addition, multiplication, subtraction, or subtraction, but will still involve those normal operations; Here is an example:

Let's define a new operation by the symbol * such that $a*b=\frac{ab}{2}$ where ab are multiplied by normal multiplication and divided by 2 under normal division. For example, $3*2=\frac{(3)(2)}{2}=3$. Now let's determine whether the closure, commutative, associative, or distributive properties hold.

Commutative: Does $a*b=b*a$? Well $a*b=\frac{ab}{2}$ and $b*a=\frac{ba}{2}$. Does $\frac{ab}{2}=\frac{ba}{2}$? Yes it does because we stated that ab are multiplied by normal multiplication and normal multiplication has commutativity $(ab=ba)$. Therefore our new operation has commutativity.

Associative: Does $a*(b*c)=(a*b)*c$? Let's simplify the left:

$$a*(b*c) => a*\left(\frac{bc}{2}\right) => \frac{a\left(\frac{bc}{2}\right)}{2} => \frac{\frac{abc}{2}}{2} => \frac{abc}{4}$$

Let's simplify the right:

$$(a*b)*c => \frac{ab}{2}*c => \frac{\left(\frac{ab}{2}\right)c}{2} => \frac{\frac{abc}{2}}{2} => \frac{abc}{4}$$

Since the left equals the right, our new operation holds under the associative property.

Distributive property: Does $a*(b+c)=a*b+a*c$?

Let's simplify the left first:

$$a*(b+c) => \frac{a(b+c)}{2} => \frac{ab+bc}{2}$$

Let's simplify the right:

$$a*b+a*c => \frac{ab}{2}+\frac{ac}{2} => \frac{ab+ac}{2}$$

Since $\frac{ab+bc}{2} \neq \frac{ab+ac}{2}$, our new operation does NOT have the distributive property.

To examine the concept of closure with our new operation, let's define a new set as the set of all odd integers and see if the set of all odd integers is closed under our new operation. In other words, given any two odd integers, $a\,and\,b$, does $a*b$ result in an odd integer? While we could easily write a formal proof of this by using $2n+1$ and $2n+3$ as the two odd integers, let's just try a couple examples:

Let 3 be our first odd integer and let 5 be our other odd integer: does $3*5$ result in an odd integer?

$$3*5 => \frac{(3)(5)}{2} => \frac{15}{2} = 7.5$$

Since 7.5 is not an odd integer, we have already found a counter example which proves that the set of odd integers is NOT closed under our new operation. Note that if $3*5$ did result in an odd integer, I would still try a few more combinations of odd integers to make sure that our set is closed under our new operation.

Example 4.11.a: Define a new operation "&" such that $a\&b = a^2b^2$ where the a^2 and the b^2 are multiplied using normal multiplication. Which of the following must be true if the new operation, &, is closed under the associative property?

A. $a\&b\&c = a\&(b\&c)$

B. $(a\&b)\&c = a\&(b\&c)$

C. $a(b\&c) = (a\&b)c$

D. $a(b\&c) = a\&b + a\&c$

Solutions to Example Problems:

Solution to 4.11.a:

The answer is B. Answer B represents a, b, and c being associated in two different ways with the same operation. Answer C may look correct, but answer C involves both normal multiplication and our new operation. Answer D has a mixture of normal multiplication, normal addition, and our new operation and looks sort of like the distributive property. In answer A on the left, there are no associated parts.

4.12. Additive Inverses and Multiplicative Inverses

If you take a number and you multiply it by its **multiplicative inverse** then you get 1. For example, since $5\left(\frac{1}{5}\right)=1$, 5 and $\frac{1}{5}$ are multiplicative inverses of each other.

If you take a number and add it to its **additive inverse**, you get zero. For example, since $3+(-3)=0$, 3 and -3 are additive inverses of each other.

Example 4.12.a: Which answer below has no multiplicative inverse?

A. 1
B. a^2
C. 0
D. Every number has a multiplicative inverse.

Example 4.12.b: What is the additive inverse of: $-a^2+4b-c$?

A. 1
B. $a^2-4b-c+1$
C. $\dfrac{1}{-a^2+4b-c}$
D. a^2-4b+c

Solutions to Example Problems:

Solution to 4.12.a:

The answer is C. Zero does not have a multiplicative inverse because no number exists that you can multiply by zero to get 1.

Solution to 4.12.b:

The answer is D. If you take answer D and add it to $-a^2 + 4b - c$, you get 0.

4.13. Roots, Powers, and Infinite Decimal Expansions

The relationship between roots and powers can best be explained through example: $x^{2/3}$ is the same as $\sqrt[3]{x^2}$; $\sqrt[7]{x}$ is the same as $x^{\frac{1}{7}}$. Infinite decimal expansions refers to the fact that 1=.99999… where the "…" means that those 9's go on forever; this can be proven through geometric series (see section 9.6 for practice with geometric series).

Here are the few rules when it comes to finding powers and roots:

- $\sqrt[m]{x^n} = x^{\frac{n}{m}}$. For example, $\sqrt[7]{x^2} = x^{\frac{2}{7}}$
- $x^m x^n = x^{m+n}$. For example, $x^2 x^3 = x^{2+3} = x^5$.
- $\dfrac{x^m}{x^n} = x^{m-n}$. For example, $\dfrac{x^5}{x^3} = x^{5-3} = x^2$.
- $(x^n)^m = x^{mn}$. For example, $(x^3)^5 = x^{15}$

Example 4.13.a: Which of the following is equivalent to $\sqrt[3]{7} \cdot \sqrt[5]{7}$?

A. integers $7^{8/15}$
B. $7^{1/15}$
C. $\sqrt[8]{49}$
D. $\sqrt[4]{14}$

Example 4.13.b: Which of the following is equivalent to $\left(x^{-\frac{2}{3}}\right)^{\frac{2}{3}}$?

A. 1

B. $\dfrac{1}{\sqrt[9]{x^4}}$.

C. $x^{\frac{4}{3}}$

D. $\dfrac{1}{x^{-\frac{4}{6}}}$

Solution to Example Problems:

Solution to 4.13.a:

The answer is A. $\sqrt[3]{7} = 7^{\frac{1}{3}}$ *and* $\sqrt[5]{7} = 7^{\frac{1}{5}}$. Thus, when we multiply $(7^{\frac{1}{3}})(7^{\frac{1}{5}})$, we add the exponents (note that we can only add the exponents because the bases, 7 and 7, are the same). When I added the exponents, I found a common denominator between 1/3 and 1/5. I get $7^{\frac{8}{15}}$ *or* $\sqrt[15]{7^8}$.

Solution to 4.13.b:

The answer is B. When you have an exponent to an exponent, you multiply them. Thus, $\left(x^{-\frac{2}{3}}\right)^{\frac{2}{3}} = x^{\frac{-2}{3}\times\frac{2}{3}}$ which is equal to $x^{-\frac{4}{9}}$. This is equivalent to $\frac{1}{x^{\frac{4}{9}}}$ which is equivalent to $\frac{1}{\sqrt[9]{x^4}}$.

4.14. Complex Numbers

The main concept that you have to remember with complex numbers is that $i = \sqrt{-1}$ *and* $i^2 = -1$. Also, every complex number has a conjugate pair. Note: I included another imaginary number problem in the practice test in chapter 10.

Example 4.14.a: If a complex number is represented by 'A', which letter represents its conjugate pair?

A. A
B. B
C. C
D. D

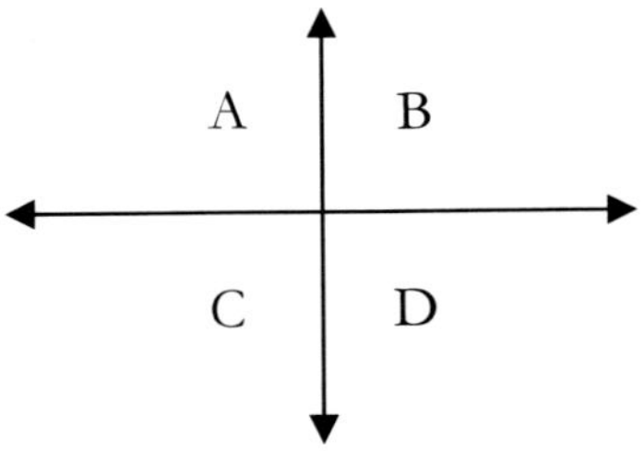

Example 4.14.b: Which of the following is an irrational number?

A. The length of a hypotenuse of a triangle which has side lengths 2 and 2.

B. $4^{\frac{1}{2}}$

C. $\frac{7}{3}$

D. $\sqrt{1}$

Solutions to Example Problems:

Solution to 4.14.a:

The answer is C. All complex conjugate pairs are mirrored across the x-axis.

Solution to 4.14.b:

The answer is A. The hypotenuse of a triangle with side lengths of 2 would be $\sqrt{8}$ (I used the Pythagorean Theorem to find that value). $\sqrt{8}$ is the only value that CANNOT be written as a fraction. In answer B, $4^{\frac{1}{2}}$ is actually $\sqrt{4}$ which is 2 or -2. In answer C, $\frac{7}{3}$ can be written as a fraction because it already is written as a fraction. In answer D, the square root of 1 is 1 or -1which can be written as the fractions: $\frac{1}{1}$ and $-\frac{1}{1}$.

4.15. Prime Factorization and Modular Arithmetic

Know your prime numbers well. When you find a least common denominator, when you reduce radicals, or when you reduce fractions, each can be done by breaking a number into all primes first. For example, to reduce the radical $\sqrt{644}$, first you can break the radiacal into primes:

$$\sqrt{644} = \sqrt{2 * 322} = \sqrt{2 * 2 * 161} = \sqrt{2 * 2 * 7 * 23} = 2\sqrt{161}$$

When I did this example, I had no idea if 161 had any factors other than 1 and 161. To figure out what factors 161 had, I divided 161 by prime numbers starting at 1, then 2, then 3, 5, 7, 11, 13, … and I figured out that 161 was divisible by 7.

Modular Arithmetic refers to concepts like clock arithmetic where you may have to add hours on a clock or fractions of a day to a clock and determine the time.

Example 4.15.a: If the number 1357913579 repeats over and over so that 1 is the 1st and 6th digits, 3 is the 2nd and 7th digits, etc., what will be the 84th digit?

A. 7
B. 1
C. 3
D. 5

Solutions to Example Problems:

Solution to 4.15.a:

The answer is A. We divide 84 by 5 because there are 5 digits in the repeating pattern and get 16 with a remainder of 4. With a remainder of 4, we take the 4th digit in the series which is 7.

Another way of looking at this problem is that the 10th, 20th, … , and 80th digit will be all be 9. Therefore the 81st digit is 1; the 82nd digit is 3; the 83rd digit is 5 and the 84th digit is 7.

4.16. Matrices

Below I have given a couple examples of questions that may arise based on matrix algebra. The examples below should shine some light on a couple of concepts in matrix algebra- specifically, matrix inverses and **matrix multiplication**. If you have trouble with matrix algebra, I suggest you find yourself an algebra II book and start doing problems. Below are some basic matrix algebra rules.

Matrix Multiplication:

When performing matrix multiplication, use the following example as a guide:

Let matrix A = $\begin{bmatrix} a & b \\ c & d \end{bmatrix}$; let matrix B = $\begin{bmatrix} 1 & 2 \\ 3 & 4 \end{bmatrix}$.

Then AB = $\begin{bmatrix} 1a+3b & 2a+4b \\ 1c+3d & 2c+4d \end{bmatrix}$. Basically, you multiply the first row of A by the first column of B to get the upper left entry in AB; multiply the first row in A by the second column in B to get the upper right entry in AB and so on.

Note: AB does not equal BA in the above example, which demonstrates that **order of multiplication in matrix multiplication is vitally important**.

Matrices larger than 2x2 can be multiplied by applying the same formula where the first row of the first matrix is compared with the first column of the second matrix to find the resulting first-row-first-column entry, etc. This procedure takes practice.

Not all matrices can be multiplied. This fact is best explained through example. Let A be a 3x2 matrix. Let B be a 2x4 matrix. To determine if AB or BA can be multiplied, write the row x column numbers next to each other like so in a little table:

A	x	B
3x2		2x4

If the inside numbers, in this case 2 and 2 are the same, then the matrices can be multiplied. The outside numbers, in this case 3x4, will be the size of the multiplied matrix AB. Thus, AB can be multiplied. Can the matrix BA be multiplied and, if so, what will the size of the resulting matrix be?

The answer is no, because if you set up the same table:

B x A

2x4 x 3x2

and 3 does not equal 4.

In the example below, I ask for you to find the inverse of a matrix. An inverse matrix is the matrix which, when multiplied by your original matrix, yields the identity matrix: $\begin{bmatrix} 1 & 0 & 0 \\ 0 & \ddots & \\ 0 & 0 & 1 \end{bmatrix}$. Note that the identity matrix can have an infinite number of rows or columns so long as every number on the main diagonal (top left to bottom right) is 1 and every other entry in the matrix is 0. The 2x2 identity matrix is: $\begin{bmatrix} 1 & 0 \\ 0 & 1 \end{bmatrix}$.

An inverse matrix is found by using any of 3 elementary row operations in the same order to your original matrix as you use them on the identity matrix. When you have turned your original matrix into the identity matrix using only **elementary row operations**, you will have simultaneously transformed your identity matrix into the inverse matrix you are looking for. Here is an example:

Say I want to find the inverse of matrix $\begin{bmatrix} 0 & 1 \\ 2 & 3 \end{bmatrix}$. First, I arrange my matrix next to the identity matrix:

$$\begin{bmatrix} 0 & 1 \\ 2 & 3 \end{bmatrix}\begin{bmatrix} 1 & 0 \\ 0 & 1 \end{bmatrix}$$

Then I perform elementary row operations until the 2x2 matrix on the left looks like the identity matrix. First, switch rows 1 and 2:

$$\begin{bmatrix} 2 & 3 \\ 0 & 1 \end{bmatrix}\begin{bmatrix} 0 & 1 \\ 1 & 0 \end{bmatrix}$$

Next, add -3 times the bottom row to the top row:

$$\begin{bmatrix} 2 & 0 \\ 0 & 1 \end{bmatrix}\begin{bmatrix} -3 & 1 \\ 1 & 0 \end{bmatrix}$$

Last, divide the top row by 2:

$$\begin{bmatrix} 1 & 0 \\ 0 & 1 \end{bmatrix}\begin{bmatrix} -3/2 & 1/2 \\ 1 & 0 \end{bmatrix}$$

And voilà! The 2x2 matrix $\begin{bmatrix} -3/2 & 1/2 \\ 1 & 0 \end{bmatrix}$ is the inverse matrix of $\begin{bmatrix} 0 & 1 \\ 2 & 3 \end{bmatrix}$.

The 3 Elementary Row Operations Are:

1. You can multiply any row by a nonzero number.
2. You can add a multiple of any row to another row.
3. You can switch any row in a matrix with another row.

Example 4.16.a: Determine the inverse of matrix $\begin{bmatrix} 2 & 3 \\ 3 & 5 \end{bmatrix}$.

A. $\begin{bmatrix} 5 & -3 \\ -3 & 2 \end{bmatrix}$

B. $\begin{bmatrix} 1 & 0 \\ 0 & 1 \end{bmatrix}$

C. $\begin{bmatrix} 0 & 1 \\ 1 & 0 \end{bmatrix}$

D. $\begin{bmatrix} -5 & -3 \\ -3 & -2 \end{bmatrix}$

Solutions to Example Problems:

Solution to 4.16.a:

The answer is A. Note that you can solve this problem in a number of ways. I chose to solve it using elementary row operations first because this method works for 2x2, 3x3, … , nxn matrices while other methods may work only for 2x2 matrices.

Start by making a matrix which has your original matrix on the left and the identity matrix on the right:

$$\begin{bmatrix} 2 & 3 \\ 3 & 5 \end{bmatrix}\begin{bmatrix} 1 & 0 \\ 0 & 1 \end{bmatrix}$$

We try to make the matrix on the left look like the identity matrix (which is currently on the right) by using elementary row operations. First, add negative row 1 onto row 2. I will show the addition as a step just in case this concept is new to you. Remember, whatever you do to the left matrix, you also do to the right matrix.

$$\begin{bmatrix} 2 & 3 \\ -2+3 & -3+5 \end{bmatrix}\begin{bmatrix} 1 & 0 \\ -1+0 & -0+1 \end{bmatrix}$$

$$=\begin{bmatrix} 2 & 3 \\ 1 & 2 \end{bmatrix}\begin{bmatrix} 1 & 0 \\ -1 & 1 \end{bmatrix}$$

Above, I multiplied the top row by -1 and added that to row 2. I did not change the top row.

Next, add -2 times row 2 onto row 1:

$$\begin{bmatrix} -2+2 & -4+3 \\ 1 & 2 \end{bmatrix}\begin{bmatrix} 2+1 & -2+0 \\ -1 & 1 \end{bmatrix}$$

$$=\begin{bmatrix} 0 & -1 \\ 1 & 2 \end{bmatrix}\begin{bmatrix} 3 & -2 \\ -1 & 1 \end{bmatrix}$$

Above, I multiplied the bottom row by -2 and added it to row 1. I did not change the bottom row.

Next, add 2 times row 1 onto row 2:

$$\begin{bmatrix} 0 & -1 \\ 1+0 & -2+2 \end{bmatrix}\begin{bmatrix} 3 & -2 \\ -1+6 & 1-4 \end{bmatrix}$$

$$=\begin{bmatrix} 0 & -1 \\ 1 & 0 \end{bmatrix}\begin{bmatrix} 3 & -2 \\ 5 & -3 \end{bmatrix}$$

Above, I multiplied the top row by 2 and added that to the bottom row. I did not change the top row.

Next, switch row 1 with row 2:

$$\begin{bmatrix} 1 & 0 \\ 0 & -1 \end{bmatrix} \begin{bmatrix} 5 & -3 \\ 3 & -2 \end{bmatrix}$$

Lastly, multiply row 2 by -1:

$$\begin{bmatrix} 1 & 0 \\ 0 & 1 \end{bmatrix} \begin{bmatrix} 5 & -3 \\ -3 & 2 \end{bmatrix}$$

Since we have made the left hand matrix into the identity matrix and because we performed all the exact same elementary row operations to the right hand matrix that we performed to the left hand matrix, the right hand matrix has become the our inverse matrix.

Alternatively, you could also work backwards by multiplying each of the matrices given as answer choices with the original matrix. If you multiply the original matrix times answer choice A (in that order), you get the identity matrix. This means that answer A is the inverse of the original matrix:

$$\begin{bmatrix} 2 & 3 \\ 3 & 5 \end{bmatrix} x \begin{bmatrix} 5 & -3 \\ -3 & 2 \end{bmatrix} = \begin{bmatrix} 2x5 + -3x3 & -3x2 + 3x2 \\ 3x5 + -3x5 & -3x3 + 5x2 \end{bmatrix} = \begin{bmatrix} 1 & 0 \\ 0 & 1 \end{bmatrix}$$

4.17. Using Matrices as Tools for Displaying Data

You have probably seen matrices represent data in the form of sports scores and menus, but matrices can be used to represent any set of data. Generally in math, data is sorted by column vectors – columns of data where every entry is a sample, but row vectors can just as easily be used.

Example 4.17.a: Doni challenged Luke to a donkey race. In this donkey race, Doni and Luke both mounted atop Donkeys, raced 100 feet to a barrel, turned around the barrel, and then raced back to the starting line. The row vectors below represent Doni's and Luke's distance from the starting line every 10 seconds. Doni's data is in the first row and Luke's data is in the second row. At how many times was Doni winning the race?

$$\begin{pmatrix} 0 & 10 & 20 & 35 & 50 & 63 & 80 & 99 & 84 & 61 & 72 & 60 & 50 & 32 & 15 & 8 & 0 \\ 0 & 11 & 22 & 38 & 52 & 61 & 87 & 98 & 85 & 60 & 50 & 41 & 31 & 14 & 3 & 0 & 0 \end{pmatrix}$$

A. 2
B. 9
C. 4
D. 3

Solutions to Example Problems:

Solution to 4.17.a:

The answer is D. Doni started behind in the race, caught up, was ahead for 3 time points (63 vs. 61, 99 vs. 98 and 84 vs. 85). Note that on the way back, a smaller number means that the donkey racer was in the lead. After that point, Doni's donkey went in the wrong direction and never caught up. I can tell you from personal experience that donkeys DO go in the wrong direction in donkey-races, and people do laugh. And it's not particularly funny for those people in the race!

4.18. Groups, Rings, and Fields

Groups, rings, and fields are subjects of abstract algebra and can get complicated to someone who has not taken an entire course. For the Praxis II ® (0061) exam, know that groups, rings, and fields are subject to the same properties of real numbers as are individual real numbers (e.g. the associative and commutative properties from earlier in the chapter). Also, groups, rings and fields are commonly used in factorization. For more on this, look up the Euclidean Algorithm.

Before we get started with groups, rings and fields, we need to review the most common sets in abstract algebra: sets of $\mathbb{Z}_n$; and we need to review the most common operations of abstract algebra: **modular arithmetic**.

$\mathbb{Z}_n$ is the set of all integers from 0 to n-1. For example, $\mathbb{Z}_7=\{0,1,2,3,4,5,6\}$ and $\mathbb{Z}_4=\{0,1,2,3\}$.

Modular arithmetic is like clock math. To do modular arithmetic, visualize a clock when you add and multiply. For example, say we are adding 5+5 in the set $\mathbb{Z}_7$. To do this, visualize a clock with seven digits (0 through 6). If you start with 5 and add another 5, you get all the way around the clock once with 3 leftover. Therefore the answer is 3. For addition and multiplication in modular arithmetic, you are actually looking for a remainder when your sum or product is divided by n.

As another example, multiply 3 times 4 in $\mathbb{Z}_5$: first you get 12. When you divide 12 by n (n=5) you get a remainder of 2. Therefore the answer is 2. If that was confusing, think of it like a clock with only 5 numbers (0 through 4). When you multiply 3 times 4 you get 12 so you go around the clock twice with a remainder of 2 hours. The answer is 2.

Here are some basics about groups, rings and fields:

A **group** is a set of numbers together with a binary operation (usually either addition or multiplication) where three properties apply:

1) The elements in a group are associative.
2) Each group has an identity, called e. If the operation in the group is addition, then e=0. If the operation is in the group is multiplication, then e = 1.
3) Every element in a group has an inverse.

Let's look at an example question: Is ($\mathbb{Z}_6$, +) a group?

First, $\mathbb{Z}_6$ is the set {0, 1, 2, 3, 4, 5} and it has the operation of addition (because of the plus sign). To determine if ($\mathbb{Z}_6$, +) is a group, we must first answer the question: are all the elements in the group associative? Well, 0+(1+2)=(0+1)+2 so that example of elements is associative; and 2+(3+4) = (2+3)+4 so that example is also associative. Continuing in this manner until you test every combination of elements, you should be able to see that ($\mathbb{Z}_6$, +) is associative.

Next, does the group ($\mathbb{Z}_6$, +) have an identity? An identity is an element such that if you add any number to the identity, you get the same element. For ($\mathbb{Z}_6$, +), the identity e=0. As an example, 0+e=0, 1+e=1, and 2+e=2, etc. if e=0. And since 0 belongs to the set $\mathbb{Z}_6$= {0,1,2,3,4,5) this implies that ($\mathbb{Z}_6$, +) has an identity.

Next, does every element in the group have an inverse? An inverse is a number such that if you take an element of the group and you add another number, you get 0. (Note, if we had multiplication as an operation rather than addition, then the inverse would be defined as the number we multiply by to get 1). Let's test for the inverse property. Does every element of ($\mathbb{Z}_6$, +) have an inverse?

0+0=0
1+5=0
2+4=0
3+3=0

And since every element has an inverse (and every inverse also belongs to our group), then our group has the inverse property.

And since all three properties hold, ($\mathbb{Z}_6$, +) is a group.

A **ring** is a set of numbers together with two binary operations (generally addition and multiplication). In addition, a ring has the following properties for $a, b, and\, c$ belonging to the ring.

1) Addition is commutative: $a+b=b+a$
2) Addition is associative: $a+(b+c)=(a+b)+c$
3) There exists a zero element such that $a+0_R=0_R+a=a$
4) Every element has an additive inverse: $a+a^{-1}=a^{-1}+a=0_R$
5) Multiplication is associative: $a(bc)=(ab)c$
6) The operations satisfy the distributive property: $a(b+c)=ab+ac$

In addition to the 6 properties above, if the following property holds then the ring is a commutative ring:

7) Multiplication is commutative: $ab=ba$

In addition, if a ring has the 7 qualities above and the property below, then the ring is a commutative ring with identity:

8) There exists an identity element such that $a*1=a$

A **field** is a commutative ring with identity with $1_R \neq 0_R$ such that for each $a \neq 0_R$ in R, the equation $ax=1_R$ has a solution. For example, the set $\mathbb{Z}_p=\{1,2,3,5,7,11,\ldots\}$ is a field where p is prime.

A few notes on notation:

$\mathbb{Z}$ stands for the set of integers unless otherwise denoted.

$\mathbb{Q}$ stands for the set of rational numbers unless otherwise denoted.

$\mathbb{R}$ stands for the set of real numbers unless otherwise denoted.

$\mathbb{C}$ stands for the set of complex numbers unless otherwise denoted.

Example 4.18.a: Which are properties of groups?

A. Groups have the associative property.
B. Every element in a group has an inverse.
C. Groups have the identity property.
D. All of the above.

Example 4.18.b: What is 5^3 in $\mathbb{Z}_4$?

A. 125
B. 0
C. 1
D. 2

Example 4.18.c: What is the multiplicative inverse of 4 in $\mathbb{Z}_9$?

A. 7
B. 1
C. 0
D. 4 has no multiplicative inverse in $\mathbb{Z}_9$.

Example 4.18.d: Find the 37th number in the following pattern where 2 is the 1st number: 2, 5, 10, 17, 26, 37, 50, …

A. 1110
B. 1212
C. 1256
D. 1370

Solutions to Example Problems:

Solution to 4.18.a:

The answer is D. All are basic properties of groups. In fact, one other basic property of groups exists - that groups possess an inverse element for each element in the set.

Solution to 4.18.b:

The Answer is C.

5 times 5 times 5 is 125, which we divide by 4 (because the question is asked in $\mathbb{Z}_4$) to get 31 with a remainder of 1. Therefore the answer is 1. Again, think of a clock. Every 4th digit is equivalent to 0: 0, 4, 8, 12, … , 120, 124. Therefore you would do 31 turns around a clock with 1 hour remaining.

Solution to 4.18.c:

The answer is A. The multiplicative inverse of 4 is the number that we have to multiply 4 by to get the number 1.

$4*x=1$. Let's check every value in $\mathbb{Z}_9$ for practice:

$4*0=0$
$4*1=4$
$4*2=8$
$4*3=12$
$4*4=16=7$
$4*5=20=2$
$4*6=24=6$
$4*7=28=1$
$4*8=32=5$

Therefore, the multiplicative inverse of 4 in $\mathbb{Z}_9$ is 7.

Solution to 4.18.d:

The answer is D. If you plot the points accurately on an x,y coordinate plane you should be able to see that the pattern is quadratic and is very close to $f(x)=x^2$. If you look at the table below and examine the changes in y, you should be able to see that the value at x=0 is y=1; thus, the pattern is one more than the number squared or $f(x)=x^2+1$.

F(x)	?	2	5	10	17	26	37	50
X	0	1	2	3	4	5	6	7

Next, I plug 37 in for x and solve for f(37): $f(37)=(37)^2+1=1370$.

You could also potentially plot the points in a TI-83 graphing calculator into the STAT -> EDIT function and then use a quadratic regression using STAT -> CALC.

Note: you absolutely need to know how to do regression analysis for the exam using the above method.

Chapter 5

Functions and Relations

5.1. Composition of Functions

As you evaluate functions of functions, work from the inside out. For example, if the function is $f(g(5))$ – read "f *of* g *of* 5" – first plug in 5's in place of every x in function $g(x)$, then plug the function g(5) into every x in function f(x).

Example 5.1.a: Find $g(f(-2))$ *where* $f(x) = x^2 + 2$ *and* $g(x) = \sqrt{x+1}$.

A. $\sqrt{7}$
B. $\sqrt{3}$
C. 1
D. 4

Solutions to Example Problems:

Solution to 5.1.a:

The answer is A. Work inside out. First plug 2 in for every x in f(x) (also denoted as f(-2)):

$$f(-2)=(-2)^2+2=4+2=6$$

Next, plug f(-2) (which we now know to be 6) into g(x) every place you see an x:

$$g(f(-2))=\sqrt{6+1}=\sqrt{7}$$

5.2. Analyzing Functions

Domain is all possible values for x. **Range** is all possible values for y. Finding range algebraically means switching x and y and then solving for y again. Finding the range algebraically can often be tricky, especially if the function is cubic or higher, but the process is relatively fast and simple if you graph the function and determine possible y values visually.

Think about range like this: how far down does the graph of the function go and how far up does the function go? – those are your lower and upper limits for your range. Likewise, how far left does your function go and how far right does your function go? Those are the lower and upper limits of your domain, respectively. See if you can figure out the domain and range below:

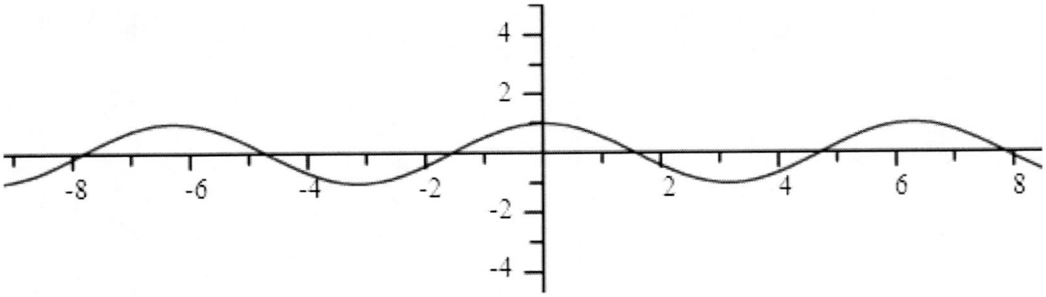

The function above goes left forever and right forever; thus the domain is $(-\infty,\infty)$. The range only goes from -1 to 1. Therefore the range is $[-1,1]$.

When determining domain and range algebraically, values that are not possible include values where the function would be divided by zero or where a negative would be in a square root. For example, the domain of $f(x)=\frac{1}{x}$ is $(-\infty,0)\cup(0,\infty)$ which includes every value except for 0 in the domain (and clearly if x is equal to 0 then we would divide by 0).

A function is **even** if it is symmetric about the y-axis (e.g. $y=x^2$). A function is **odd** if it is **symmetric** about the origin (e.g. $y=x^3$). There are algebraic methods for determining whether a function is even or odd (or neither), but the point here is to use methods that work *quickly and effectively.*

A **maximum point** is the point where a function stops going up and starts going down. A **minimum** is a point where a function stops going down and starts going up.

The slope of a maximum or of a minimum will be 0 which means, in calculus terms, that setting the first derivative to 0 and solving will give you all points where the slope is 0 (your maximums and minimums).

Example 5.2.a: What is the domain of the function: $f(x)=\dfrac{x^2+1}{x+3}$?

A. $(-\infty,\infty)$

B. $(-\infty,3)\cup(3,\infty)$

C. $(-\infty,-3)\cup(-3,\infty)$

D. $(-\infty,-3]\cup[3,\infty)$

Example 5.2.b: Luke starts from point A and runs as fast as he can toward point B in a straight line. As Luke approaches point B, he slows down and eventually stops at point B. Luke then turns around and runs back as fast as he can toward point A in a straight line. As Luke approaches point A, he slows down and eventually stops at point A. If a Luke's distance from point A is graphed on the y-axis and time is graphed on the x-axis, which will accurately reflect the maximum on the graph?

A. There are two maximums: one where Luke is running his fastest on the way to point B and one where he is running his fastest on the way back from point B.

B. When Luke returns to point A.

C. When Luke is stopped at point B.

D. When Luke is running at high highest speed, wherever that is.

Solutions to Example Problems:

Solution to 5.2.a:

The answer is C. You can tell immediately that x = -3 is not in the domain because if you plug -3 in for x, you get a 0 in the denominator of $f(x)=\frac{x^2+1}{x+3}$. As you read the answers, D looks like it may be correct but answer D is incorrect because the square brackets in answer D actually mean that -3 is included in the domain.

Solution to 5.2.b:

The answer is C. It is probably best to sketch out a graph in order to best demonstrate this problem:

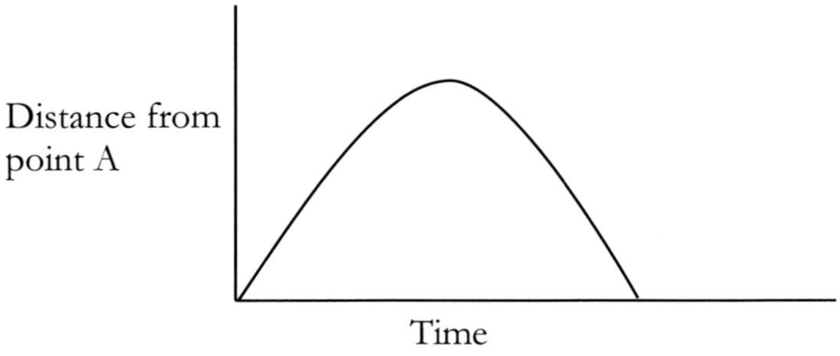

It may seem strange, but the graph is at a maximum when Luke's distance from point A is furthest – that is, when he is stopped at point B.

5.3. Transforming Functions

When in doubt, plug in numbers or functions into your graphing calculator and observe the resulting transformation. This is the standard form for a parabola:

$$y=a(x-h)^2+k$$

where h translates the parabola in the x-dimension, k translates the parabola in the y-dimension, and a is the amplitude. If a>1, the parabola increases faster; if 0<a<1, the parabola increases slower; and if a<0, the parabola opens down.

Here is an example which shows the basis for most transformations:

$$y=x^2 \quad \text{versus} \quad y=-(x+2)^2+3$$

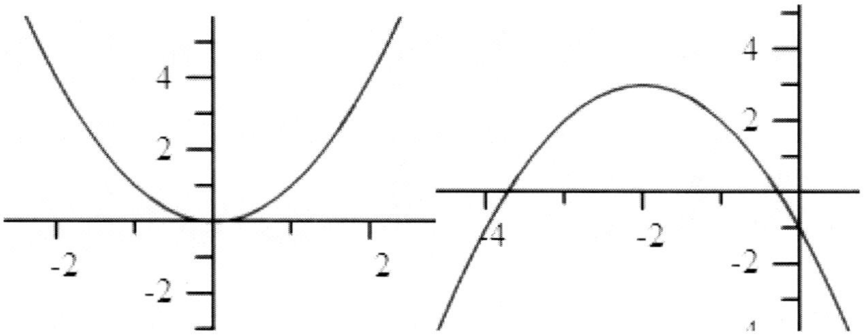

In the transformation above, the vertex of the **parabola** moves 2 units to the left (inside the parentheses) and 3 units up (outside the parentheses). For a parabola to become skinnier, fatter, or flip upside down, the "a" would change by becoming greater than 1, a fraction, or negative respectively (e.g. $y = \frac{-1}{2}x^2$ is a fatter, upside down parabola).

This same system of transformation applies to trigonometric functions, **cubic** functions, **conic** sections and more.

Example 5.3.a: Use the graphs of $f(x)$ and $g(x)$ to answer the following question. Which graph *best* represents $(f+g)(x)$?

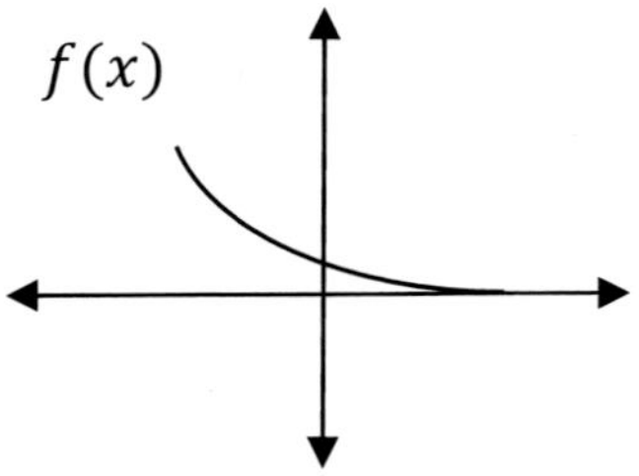

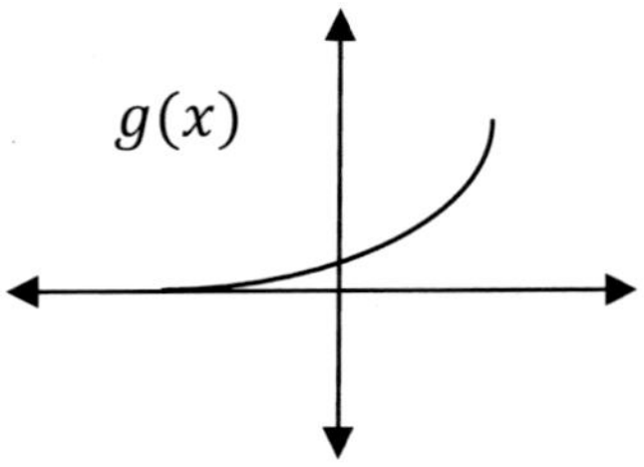

A.

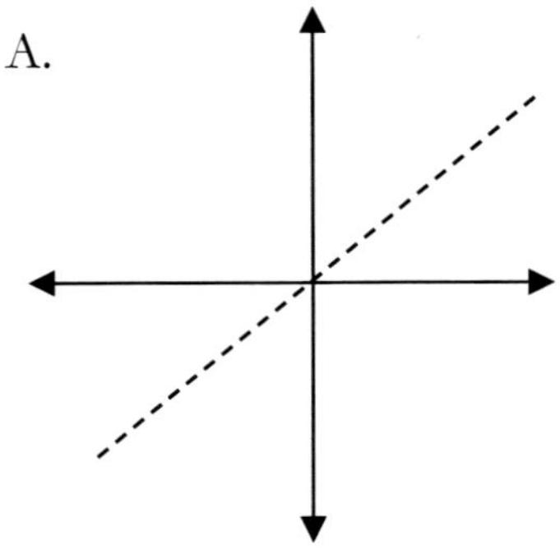

B.

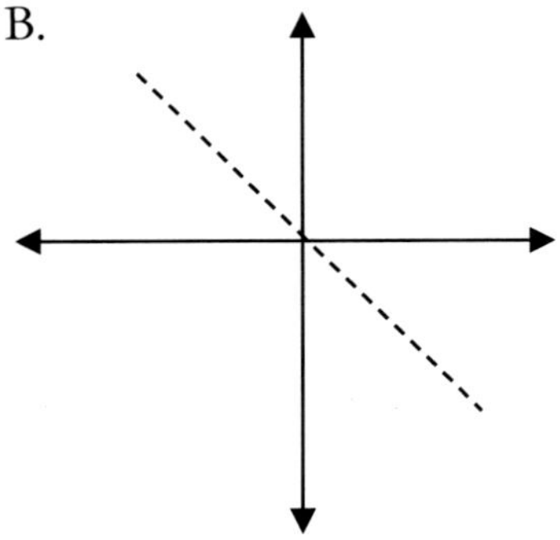

C.

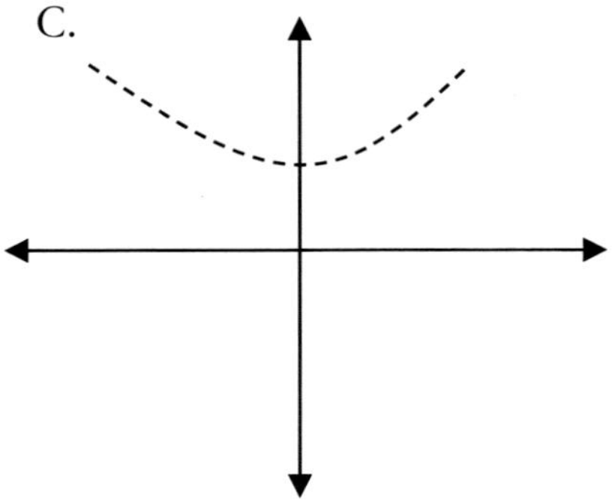

D

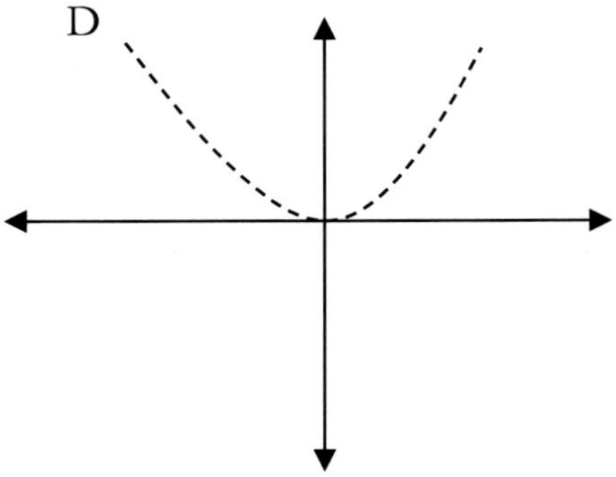

Solutions to Example Problems:

Solution to 5.3.a:

The answer is C. To take $(f+g)(x)$, you add the y values from each function. If you get to a problem such as this on the examination, do a quick experiment on your graphing calculator; type in 3 functions: $f(x)=x^2; g(x)=2; and\ (f+g)(x)=x^2+2$. The third of these functions is a parabola raised up 2 units and demonstrates that the answer will represent an addition of y-values at each x value of the original functions.

5.4. Finding Inverses

To find an inverse of a function, switch x and y and solve for y again.

Example 5.4.a: What is the inverse of $y=\sqrt{x+3}$?

A. $y=x^2-3; x\geq -3$

B. $y=\sqrt{(x+3)}; x\geq -3$

C. $x=\sqrt{y+3}; y\geq -3$

D. $x=y^2-3; y\geq 0$

Solutions to Example Problems:

Solution to 5.4.a:

The answer is A. To find an inverse, switch x and y and solve for y again. Thus,

$x = \sqrt{y+3} \Rightarrow x^2 = y+3 \Rightarrow y = x^2 - 3$. Thus, $y = x^2 - 3$ is the inverse.

After you find the inverse of a function, you can determine the range of the function. When you find the range of the function, every value will be in the range EXCEPT those values which give you a 0 in a denominator or a negative in a square root.

In the above example, there aren't any numbers that you can put for x in $y = x^2 - 3$ that result in a negative inside of a square root (there is no square root) nor are there any numbers you can put in for x in $y = x^2 - 3$ that result in a 0 in the denominator, which means that the range of the original function is $(-\infty, \infty)$.

However, the domain of the original function $y = \sqrt{x+3}$ is limited because any number below 3 that you put in place of x results in a negative within a square root. Thus, the domain of the function $y = \sqrt{x+3}$ is $(-3, \infty)$.

5.5. Representing Functions

Functions can be represented in many ways. The key here is to understand that a function is merely a system for inputs and outputs. The example below connects two functions algebraically through a geometric model.

Example 5.5.a: A square is inscribed in a circle as shown below. Write a formula for the **area of a circle**, A, as a function of the side length of the square, s.

A. $A = \pi s^2$

B. $A = \pi^2 s$

C. $A = \pi(\frac{s^2}{2})$

D. $A = \pi\left(\frac{s}{2}\right)^2$

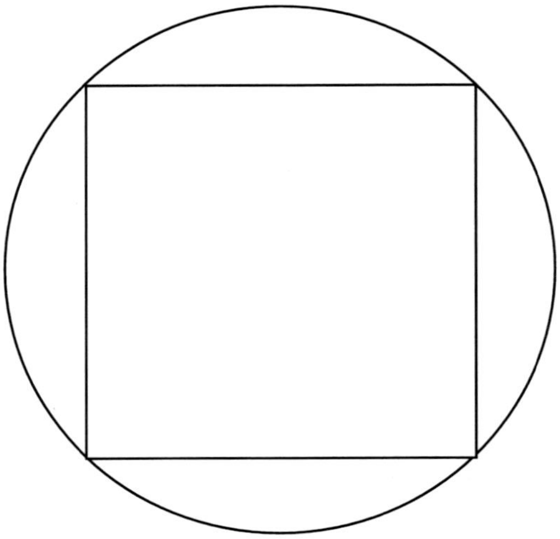

Solutions to Example Problems:

Solution to 5.5.a:

The answer is C. Begin by drawing a diagonal between any two corners of the square (as shown below); in drawing a diagonal, you create a way to compare variables of the square and the circle. The line that connects the two corners of the square is the **diameter** of the circle.

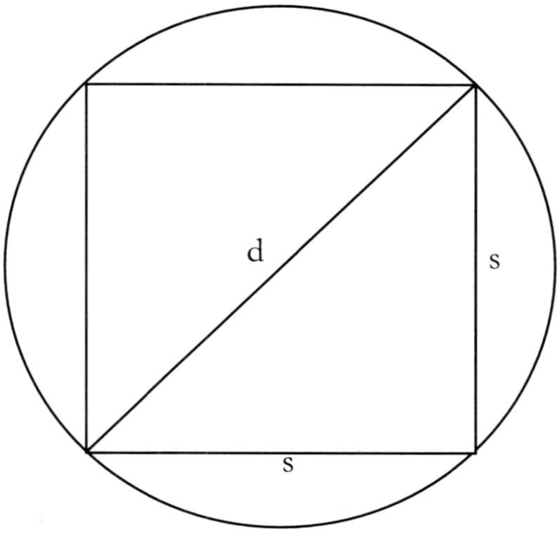

At this point, you can proceed in a number of ways, but the end goal is to equate the **radius of the circle** with the side length of the square.

I start with two equations and then try to combine them by substitution:

$$s^2 + s^2 = d^2$$

d=2r (r is the radius, d is the diameter, and s is the side length)

Thus, $s^2 + s^2 = d^2$ or $2s^2 = d^2$. Now, since d = 2r we can substitute r in for d:
$2s^2 = (2r)^2 \Rightarrow 2s^2 = 4r^2 \Rightarrow s^2 = 2r^2$. Solving for r so that we can substitute s in for r, we get:
$\frac{s^2}{2} = r^2 \Rightarrow \frac{s}{\sqrt{2}} = r$. Now we substitute $\frac{s}{\sqrt{2}}$ in for r in the equation for the area of a circle:

$$A = \pi \left(\frac{s}{\sqrt{2}} \right)^2 = \pi \frac{s^2}{2}$$

5.6. Linear Functions

The Praxis II ® (0061) exam creators will throw tricky questions at you to make linear functions more difficult. A tabular representation of linear functions would include a table, a graphic representation of linear functions would include a graph or a series of graphs and an algebraic representation of linear functions would give you an expression or many expressions. Below is an example of a verbal representation of linear functions.

Example 5.6.a: Function A is linear and crosses the x- axis only one time at $x=-2$. Function A crosses the y-axis at $y=4$. Function B is perpendicular to Function A. Which of the following equations could represent
Function B?

A. $y=2x+5$

B. $y=-\frac{1}{2}x+24$

C. $y=-\frac{1}{2}x+1$

D. Two of the above

Solutions to Example Problems:

Solution to 5.6.a:

The answer is D. Draw the figure out (as shown below). To begin with, function A is a straight line because it is linear. If you plot the function, you can tell that the slope is +2.

This means that any line which is perpendicular to function A must have a slope that is the negative inverse of +2; thus, the slope of function B must be -1/2. Therefore, you can use process of elimination and cross answer A out. Since answers B and C can each be perpendicular to function A, the answer is two of the above.

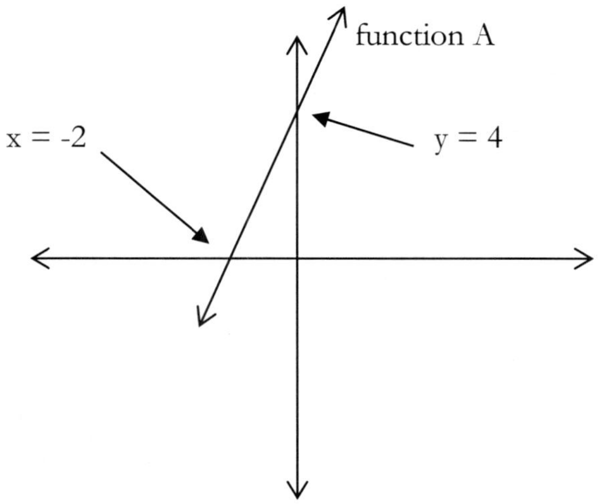

5.7. Linear Equations

Linear functions are functions for straight lines. $y = mx + b$ is an effective format for interpreting where a function lies on the x-y plane and what its slope, *m*, is. *b* is the y-intercept.

When working with linear functions, plug in coordinates into your $y = mx + b$ until all the pieces work. Remember that slope, *m*, is simply rise over run or $\frac{change\,in\,y}{change\,in\,x}$.

Example 5.7.a: A line goes through the points $(2,5)$ and $(-1,3)$. What is the equation for the line?

A. $y=2/3x+11/3$
B. $y=2/3x-11/3$
C. $y=3x+2$
D. $y=-3/2x+11/3$

Example 5.7.b: Given that the equation for a line is $Ax + By = C$ where A, B, and C are non-zero, what is the x-intercept of the line?

A. $\frac{C}{A}$

B. $\frac{A}{C}$

C. $-\frac{A}{B}$

D. Not enough information given.

:Solutions to Example Problems:

Solution to 5.7.a:

The answer is A. Start with the slope: the rise of the line is $\frac{rise}{run} = \frac{3-5}{-1-2} = \frac{-2}{-3} = \frac{2}{3}$. So far, we know that $y = \frac{2}{3}x + b$. Next, we plug in one of the coordinate points; let's use (2, 5):

$$(5) = \frac{2}{3}(2) + b \Rightarrow 5 = \frac{4}{3} + b \Rightarrow b = \frac{11}{3}$$

Thus, we know that the equation for our line is $y = \frac{2}{3}x + \frac{11}{3}$.

Solution to 5.7.b:

The answer is A. The x-intercept is the place where the function crosses the x-axis. If you realize that the function will cross the x-axis when $y = 0$, then we can simply plug in 0 for y and solve for x:

$$Ax + B(0) = C$$

$$Ax = C$$

$$x = \frac{C}{A}$$

5.8. Systems of Linear Equations and Inequalities

The example below tests your knowledge of systems of linear equations. Try to solve it by either:

1) solving for a variable and then substituting into the other equation

2) lining up the two equations, multiplying one of the equations by a constant and adding the two equations together

3) converting the system of inequalities to matrix form and performing elementary row operations on them

Example 5.8.a: Evaluate:

$$3x+4y=1$$
$$2x-3y=4$$

A. $y=-\frac{10}{19};x=\frac{51}{112}$

B. $y=-\frac{10}{17};x=\frac{19}{17}$.

C. $y=-\frac{5}{8};x=\frac{5}{8}$

D. $y=-\frac{3}{12};x=\frac{55}{111}$

Example 5.8.b: Which system of linear inequalities is represented by the graph below?

A. $y\geq\frac{7}{3}x+2 \quad y\geq-\frac{3}{4}x+2$

B. $y>\frac{7}{3}x+2 \quad y=-\frac{3}{4}x+2$

C. $y>\frac{8}{3}x+2 \quad y\geq-\frac{3}{4}x+2$

D. $y<\frac{8}{3}x+2 \quad y\leq-\frac{3}{4}x+2$

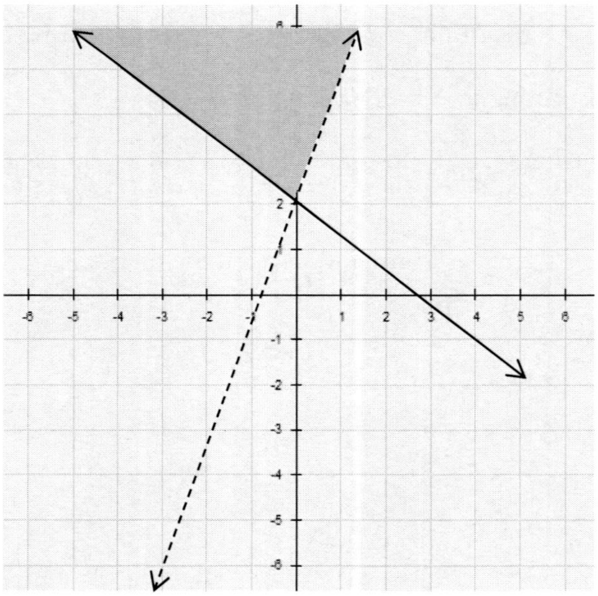

Solutions to Example Problems:

Solution to 5.8.a:

The answer is B. I am going to start this by multiplying the top equation by -2 and multiplying the bottom equation by 3:

$$-6x-8y=-2$$

$$6x-9y=12$$

Now I add the two equations together which cancels out the x's and allows us to solve for y:

$$-17y=10$$

$$thus, y=-\frac{10}{17}$$

Now, we plug $-\frac{10}{17}$ in for y into either equation and solve for x:

$$3x+4\left(-\frac{10}{17}\right)=1$$

$$\Rightarrow 3x-\frac{40}{17}=1 \Rightarrow 3x=\frac{57}{17} \Rightarrow x=\frac{57}{51} \Rightarrow x=\frac{19}{17}$$

Note: you can work backwards on this problem by simply plugging in both answer choices; if both equations work with the answer choices then the answer choice is correct.

Solution to 5.8.b:

The answer is C. To solve this problem, we first need to figure out the equation for each line:

The dotted line has a slope of $m=\frac{8}{3}$ and a y-intercept of 2; I found the slope by finding two easily recognizable points on the dotted line: (-3,-6) and (0,2) and counting the $\frac{rise}{run}$ to get my slope. Thus, the equation for the dotted line is $y=\frac{8}{3}x+2$.

The solid line goes through the points (0,2) and (4,-1). Therefore it has a slope of $m=-\frac{3}{4}$ and a y-intercept of 2. Thus, the equation of the solid line is $y=-\frac{3}{4}x+2$.

However, the graphed region is strictly above the dashed line (because it is dashed). We represent this by saying that $y>\frac{8}{3}x+2$. The graphed region is on AND above the solid line; thus $y\geq-\frac{3}{4}x+2$.

5.9. Real-World Linear Equations and Inequalities

You can problem solve in many different ways; try tables, try graphing, try creating algebraic equations. In the example below, try to create algebraic equations for the two brothers.

Example 5.9.a: Luke starts with \$200 and donates \$10 per day to the homeless shelter. His brother, Doni, starts with \$140 and donates \$5 per day. When does Doni have more money left than Luke?

A. day 12 and after
B. day 13 and after
C. day 14 and after
D. day 15 and after

Solutions to Example Problems:

Solution to 5.9.a:

The answer is B. You could make a table, but the quickest way to solve this problem is by creating an algebraic inequality. Luke's money is represented by $200-10x$ where x is the amount of days. Doni's money is represented by $140-5x$. What we want to know is when Luke's money < Doni's money. Thus,

$$200-10x<140-5x$$

Solving one side for x, we have $60<5x or 12<x$. Be careful here. The expression $12<x$ means that for the inequality $200-10x<140-5x$ to be true, x must be GREATER than 12. The answer is day 13 and after.

5.10. Quadratic Functions: Real and Complex Roots

Factoring. Remember that if a function is quadratic, you can factor, **complete the square**, or use the quadratic formula to find its roots. You can also graph a function and look to see where the function touches the x-axis – this is where the "roots" are.

If a quadratic function (a parabola) only touches the x-axis once, the function will have a **repeated root** and will factor into $(x-h)^2$. If you graph a function and it does not touch the x-axis, then there are no real roots meaning that there are two complex roots.

Example 5.10.a: What are the roots of $f(x)=x^2+3x+5$?

A. $x=\frac{-3}{2} \pm \frac{i\sqrt{11}}{2}$

B. $x=-3,5$

C. $x=-3 \pm \frac{i\sqrt{11}}{2}$

D. $x=\frac{-3}{2} \pm \frac{i\sqrt{29}}{2}$

Solutions to Example Problems:

Solution to 5.10.a:

The answer is A. You can go about solving the function in a number of ways: the first is to graph the function to see if you get any clues, the second is to plug in answers and see which ones work (and to cancel some out this way), the third is to complete the square a fourth method, which I will use, is plugging the numbers into the quadratic formula (simply because this method is quickest). Note: a=1, b=3, and c=5 since $f(x)=ax^2+bx+c$:

$$0=\frac{-b\pm\sqrt{b^2-4ac}}{2a}=\frac{-3\pm\sqrt{3^2-4(1)(5)}}{2(1)}=\frac{-3\pm\sqrt{9-20}}{2}=\frac{-3\pm\sqrt{-11}}{2}$$

From here you can tell that the answer will be A. Note that answer C is a trick answer based on a common mistake that many people make when they do not divide the first term in the quadratic formula by 2a.

5.11. Representing Quadratic Functions

The Praxis II ® (0061) exam will test your problem solving skills, often by making you connect multiple representations of quadratic functions.

Example 5.11.a: Given $f(x)=ax^2+bx+c$, which of the following conditions yields a pair of complex roots?

A. $2a<4b$
B. $b>4ac$
C. $c<abx$
D. $4ac>b^2$

Solutions to Example Problems:

Solution to 5.11.a:

The answer is D. Examine the quadratic formula: $\frac{-b \pm \sqrt{b^2 - 4ac}}{2a}$. If $b^2 - 4ac$ (often referred to as the **determinant**) is less than 0, then you will have a pair of complex roots. Thus, $b^2 - 4ac < 0 or b^2 < 4ac$. If you flip $b^2 < 4ac$ around, you have answer D.

5.12. Solving Quadratic Equations and Inequalities

When answering questions about quadratic inequalities, think to yourself, "which x- or y-values make the inequality true?"

In the following example, try to use graphical techniques to find your answer.

Example 5.12.a: Which values of x make the expression $0 \leq x^2 - 5x + 4$ true?

A. $(-\infty, \infty)$

B. $(1,4)$

C. $(-\infty, -4] \cup [1, \infty)$

D. $(-\infty, 1] \cup [4, \infty)$

Solutions to Example Problems:

Solution to 5.12.a:

The answer is D. Graph the equation $y = x^2 - 5x + 4$ on your calculator. Essentially, this question is asking "when is $x^2 - 5x + 4$ on or above the x-axis." In other words, when are the y-values of $x^2 - 5x + 4$ greater than or equal to 0.

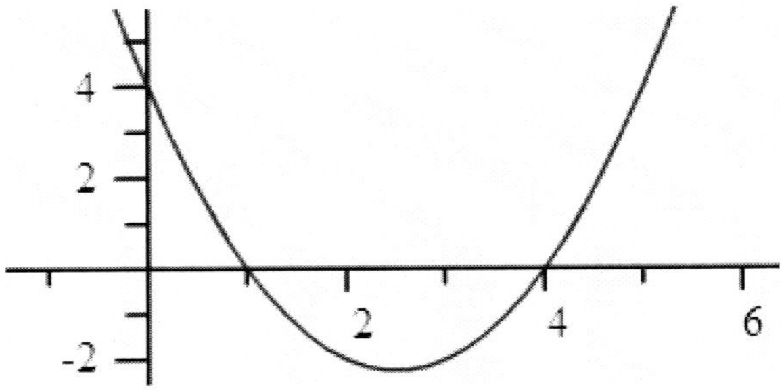

As you can see by the graph, the y-values are greater than 0 from -∞ to 1 and from 4 to ∞.

5.13. Real-World Quadratics

Most quadratic equations (and other degree equations) can be analyzed using **regression analysis** (following bullet point).

For determining which quadratic equation fits a data set, enter the data into your graphing calculator and use the quadratic regression option. On a Texas Instruments calculator, the data set can be entered under STAT -> EDIT and can be analyzed using STAT -> CALC -> QUADREG. This should give you a quadratic formula which fits the data given.

Example 5.13.a: Luke throws a Frisbee as hard as he can and his buddy, Donny, estimates the height of the Frisbee during its first 5 seconds. Assuming the Frisbee continues to follow this parabolic path (which it obviously will not), what will the height of the Frisbee be at 8 seconds?

A. 26 yards
B. 27 yards
C. 28 yards
D. 29 yards

Time(seconds)	Height (yards)
1	1
2	2
3	4
4	7
5	11

Solutions to Example Problems:

Solution to 5.13.a:

The answer is D. Since the Frisbee follows a parabolic path, you should use quadratic regression to answer the question. Plugging the coordinates into my calculator as described above and using the QuadReg command, I get the expression $y = .5x^2 - .5x + 1$. Note that x represents time in seconds, and y represents height in yards. We need to find the height at x = 8 seconds so I plug 8 in for x:

$$y = .5(8)^2 - .5(8) + 1 \Rightarrow y = .5(64) - 4 + 1 \Rightarrow y = 32 - 3 \Rightarrow y = 29\, yards$$

5.14. Polynomials: Finding Real and Complex Roots

A polynomial equation will have the same number of roots as its highest degree (for example a 3rd degree polynomial will have 3 roots – either real or complex or a combination of real and complex). Use whatever techniques you can to determine the roots. With third and higher degree polynomials, finding roots algebraically is very difficult; to find 3rd and higher degree roots effectively, graph the function and determine its roots by where it crosses the x-axis.

On the Praxis II ® (0061) exam, you should have at least one or more roots that cross the axis at a reasonable place (like at integer values); use these values to find the others. If you can divide out roots and reduce your polynomial to a quadratic polynomial, you can use the quadratic equation to factor out the remaining roots; the answer to the example problem below expands upon this technique.

Example 5.14.a: What are the real and complex roots of $f(x) = x^3 + x^2 - 5x - 6$?

A. $x = -2, -3$

B. $x = -1, \pm 1 + 3i$

C. $x = -2, \frac{1}{2} \pm \frac{\sqrt{13}}{2}$

D. $x = -2, -\frac{1}{2} \pm \frac{\sqrt{13}}{2}$

Solutions to Example Problems:

Solution to 5.14.a:

The answer is C. Start by graphing the function:

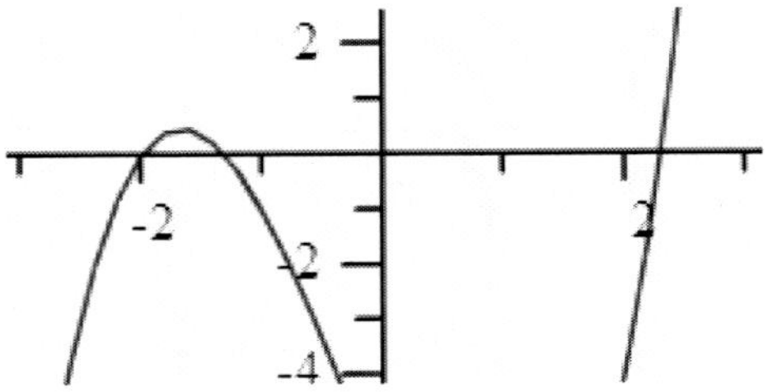

Because the function crosses the x-axis at x = -2, I can tell that -2 is the only rational root. You can check this by plugging -2 in for x in the function and seeing that the result is 0. Since -2 is the only rational root, you can use process of elimination to cancel out answer A (because -3 is clearly not a root because the function does not cross the x-axis at 3). You can cancel out answer B because the function actually crosses the x-axis 3 times and since there are 3 roots for any cubic equation, there must be 3 real roots.

To find the resulting roots, you can either work backwards (plug $\frac{1}{2}\pm\frac{\sqrt{13}}{2}$ in and see if they result in a 0) or you can do **synthetic division**. I will do synthetic division which begins by synthetically dividing x^3+x^2-5x-6 by our known root (-2):

$$\begin{array}{r|rrrr} -2 & 1 & 1 & -5 & -6 \\ & & -2 & 2 & 6 \\ \hline & 1 & -1 & -3 & 0 \end{array}$$

The 1 1 -5 -6 in the synthetic division above represents the coefficients of the function. I began the synthetic division by bringing down the 1, then multiplying the 1 by -2 and placing that number below the next 1. Then I added 1 and the -2 to make -1. Then I multiplied the -1 by the -2 and placed the 2 below the 5. Then I added the -5 and the 2 to make -3. Then I multiplied the -2 times the -2 to make 6. Then I added the -6 and the 6 to make 0. Because the end result of the synthetic division was 0, I know that -2 really is a root of x^3+x^2-5x-6.

The result of my synthetic division is $f(x)=(x+2)(x^2-x-3)$ because I successfully factored a -2 out of my original function (you can see that if you plug -2 into the term x+2, you get zero for the entire equation). The resulting coefficients 1, -1, -3, and 0 are the coefficients of the function with x+2 factored out.

Next, I can use the quadratic equation on the second term of the function x^2-x-3 to find the remaining roots:

$$0=\frac{1\pm\sqrt{1-4(1)(-3)}}{2}=\frac{1\pm\sqrt{1+12}}{2}=\frac{1}{2}\pm\frac{\sqrt{13}}{2}$$

5.15. Representing Absolute-Value, Radical, and Polynomial Functions

You should know what the absolute value function looks like (on the right) and what $f(x)=\sqrt{x}$ (on the left) looks like. The example below tests your knowledge of translating a cubic function.

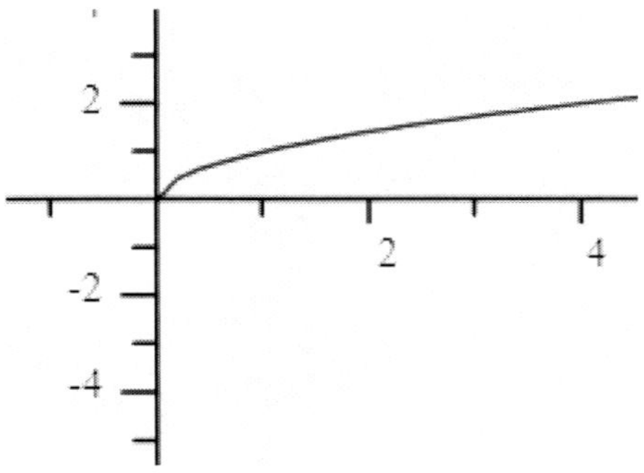

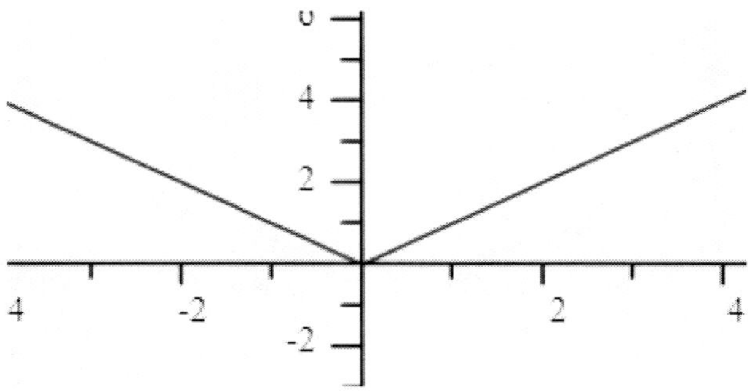

Example 5.15.a: The function $f(x)=ax^3+bx^2+cx+d$ has a mirror image about the x-axis. This mirror image is the function $g(x)$. What is $g(x)$?

A. $g(x)=ax^{-3}+bx^{-2}+c/x+d$

B. $g(x)=-ax^3-bx^2-cx-d$

C. $g(x)=-ax^3+bx^2+cx+d$

D. $g(x)=f^{-1}(x)$

Solutions to Example Problems:

Solution to 5.15.a:

The answer is B. Pick values for a, b, c, and d (I would choose all 1's) and try out the answers one at a time on your graphing calculator while graphing the original function on the same graph. Here is my result for answer B:

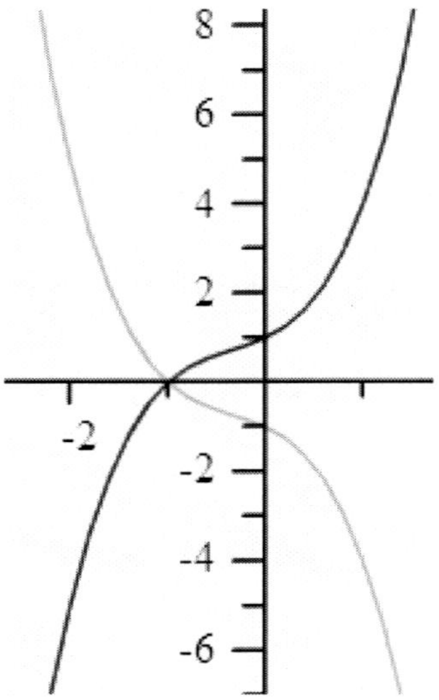

Clearly, answer B represents a mirror image about the x-axis because when we make every part of the polynomial negative, we get a perfect mirror about the x-axis.

5.16. Solving Absolute-Value, Radical, and Polynomial Functions

---

There is an interesting rule when it comes to solving absolute value inequalities. I'll demonstrate through an example that contrasts the difference between greater than and less than absolute value inequalities:

$$|x-1| \leq 5 \; vs. \; |x-1| \geq 5$$

First, the "less than":

It's straightforward. Remove the inequality, insert the negative on the left, and treat it as a single expression:

$-5 \leq x-1 \leq 5$
$+1 \qquad +1 \; +1$ Add 1 to all three parts
$-4 \leq x \leq 6$

And you're finished.

But look what happens if you do the same with a greater than expression:

$-5 \geq x-1 \geq 5$
$+1 \qquad +1 \; +1$ Add 1 to all three parts

$-4 \geq x \geq 6$ Next remove the x.
$-4 \geq 6$ And you get a false statement.

Thus, you have to treat the problem as two different statements instead of a single one:

$|x-1| \geq 5$

$x-1 \leq -5 \text{ or } x-1 \geq 5$ Add 1 to each expression.

$x \leq -4 \text{ or } x \geq 6$

Try solving the function below by drawing a number line and plugging in numbers. The algebraic steps will be discussed in the answer.

Example 5.16.a: Solve for x if:

$|x+4| \geq 3$

A. $-7 \leq x$ or $x \leq -1$

B. $-7 \geq x \geq -1$

C. $-7 \leq x \leq -1$

D. $-7 \geq x$ or $x \geq -1$

Solutions to Example Problems:

Solution to 5.16.a:

The answer is D. Remove the absolute value in $|x+4| \geq 3$ by turning the expression into two expressions where the less than expression will have a negative version of the right hand side:

$$-3 \geq x+4 \text{ or } x+4 \geq 3$$

Then solve by subtracting 4 from each expression:

$$-7 \geq x \text{ or } x \geq -1$$

You can also do this problem by working backwards (plugging in answer choices until you find one that makes the expression true).

5.17. Real-World Absolute-Value, Radical, and Polynomial Functions

---

The example below is a classic real world polynomial problem. As a hint, try creating an algebraic model for the problem using equations that you already know.

Example 5.17.a: 4 squares of side length x are cut out of the corners of an 8.5inch x 11inch piece of paper such that the sides can be folded up to form an open-topped box as shown below. What value of x makes the **maximum volume** of the box?

A. 1.85
B. 1.74
C. 1.52
D. 1.58

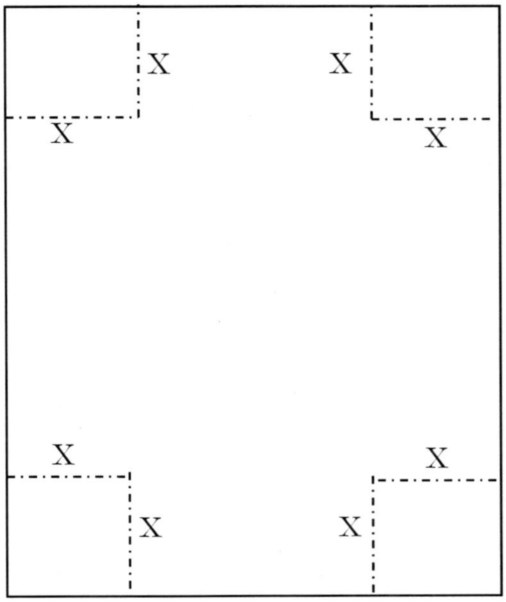

Solutions to Example Problems:

Solution to 5.17.a:

The answer is D. You will quite possibly see a maximum value problem on the exam. If you are having trouble visualizing this problem, grab a piece of paper and cut out squares from the sides and fold them up.

The key here is to realize that the length of the paper is 11-2x; the width is 8.5-2x; and the height, once folded up, will be x. Thus the volume of the open top box can be written as:

$$Volume = lenth\,x\,width\,x\,height = (11-2x)(8.5-2x)(x)$$

Plugging that equation into your graphing calculator exactly as is (no need to multiply it out), and resizing your window to accommodate the graph of the function, we get:

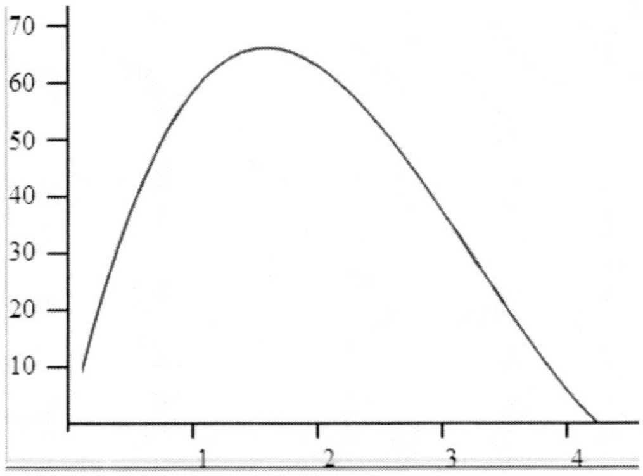

Notice the x-axis does not go further than 4.25 because we cannot cut x further than 4.25inches in from both sides; [0, 4.25] is the domain. The y-axis represents volume. using the CALC→Maximum on your graphing calculator and focusing on the peak of that function (to do this, you choose a point left of the peak, then a point right of the peak, then press enter again), you get 1.58 as the length of x which results in a maximum volume and you get 66.14 cubic inches as a maximum volume.

5.18. Exponents and Logarithms

There exist only a few essential logarithmic laws that you need to know for the Praxis II ® (0061) exam: the **change of base** formula, the quotient rule, the product rule, the power rule, basic properties of **natural logarithms**, and basic properties of common logarithms. Here they are:

The change of base formula is: $\log_a x = \dfrac{\log_b x}{\log_b a}$. Example: $\log_2 5 = \dfrac{\log_{10} 5}{\log_{10} 2} = \dfrac{.699}{.301}$. The change of base formula is useful because you can't type a log base 2 into your calculator, but a log base 10 is simply the log button on your calculator.

The quotient rule for logarithms is: $\log_a\left(\dfrac{x}{y}\right) = \log_a x - \log_a y$.

The product rule for logarithms is: $\log_a(xy) = \log_a x + \log_a y$.

And the power rule for logarithms is: $\log_a x^y = y \log_a x$.

In the example problem below, you must deal with an exponential function (exponential functions go hand-in-hand with logarithms). You cannot algebraically compare the left and right hand sides of this equation because they have different bases. Try to get the bases equal first.

Example 5.18.a: Solve for x if:

$$100\left(\frac{1}{5}\right)^{x/5} = 4$$

A. $x = 2$

B. $x = 1/20$

C. $x = 10$

D. $x = 20$

Solutions to Example Problems:

Solution to 5.18.a:

The answer is C. The bases of the equation above are 1/5 and 4. In order to compare sides, we must first get those bases equal. I start by dividing both sides by 100 to simplify the situation:

$$\frac{1}{100}(100)\left(\frac{1}{5}\right)^{\frac{x}{5}}=(4)\left(\frac{1}{100}\right)\Rightarrow\left(\frac{1}{5}\right)^{\frac{x}{5}}=\frac{1}{25}$$

Now our bases are 1/5 and 1/25. We need to make them the same. Here is where the trick of this problem comes in: realizing that 1/25 is the same as $(\frac{1}{5})^2$. Substituting this in, we get the equation:

$$\left(\frac{1}{5}\right)^{\frac{x}{5}}=\left(\frac{1}{5}\right)^{2}$$

Hopefully you can see that if $\frac{x}{5}=2$, then both sides become equal. Multiplying it out, we get x = 10. You can also work backwards in this problem by plugging in answers until the left side of the equation equals the right side.

5.19. Converting Between Exponential and Logarithmic Functions

---

The inverse relationship between logarithms and exponential functions is: $y=\log_b(x)$ if and only if $b^y=x$; this formula is, perhaps, the most important logarithm formula to memorize for the examination.

Example 5.19.a: Which of the following is equivalent to $\log_3 4 = x$?

A. $4^3 = x$

B. $x^4 = 3$

C. $3^x = 4$

D. $\log_4 3 = x$

Solutions to Example Problems:

Solution to 5.19.a:

The answer is C. I like to read $\log_3 4 = x$ like this: "three to the x power equals four." Thus, our answer is $3^x = 4$. This is one of those formulas that you just have to remember.

5.20. Trigonometry Basics

Remember **SOHCAHTOA** and how to use it, and know your 30-60-90 and 45-45-90 triangles by heart. Here they are:

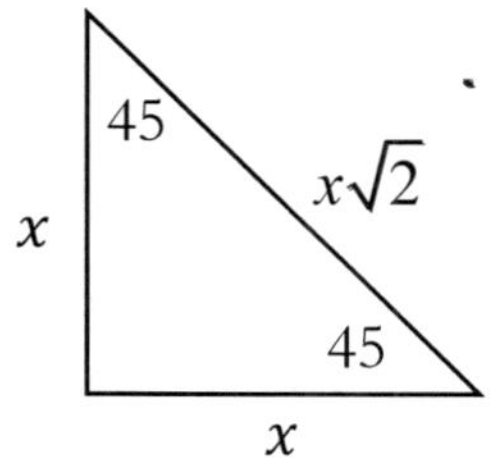

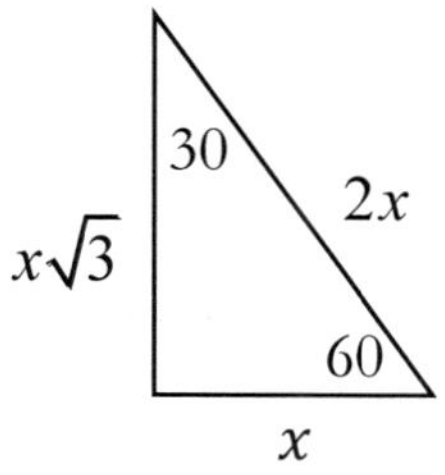

Note: there are more trig problems in the practice exam in chapter 10.

Example 5.20.a: If the tangent of $\theta = \frac{1}{\sqrt{3}}$, which represents the **sine** of θ?

A. ½

B. $-½$ *or* ½

C. $\frac{\sqrt{3}}{2}$

D. $-\frac{\sqrt{3}}{2}$ *or* $\frac{\sqrt{3}}{2}$

Solutions to Example Problems:

Solution to 5.20.a:

The answer is B. Since the tangent of theta is $\frac{1}{\sqrt{3}}$, this means that the opposite side of angle theta is 1 unit and the adjacent side of angle theta is $\sqrt{3}$. I will sketch this out below:

Note that both $\frac{1}{\sqrt{3}}$ *and* $\frac{-1}{-\sqrt{3}}$ each result in a tangent of $\frac{1}{\sqrt{3}}$ (left and right triangles, respectively).

Hopefully you can tell that these are 30-60-90 right triangles and that the hypotenuse of each is 2. Therefore:

$$sin\theta = \frac{opposite}{hypotenuse} = \frac{1}{2}\, and - \frac{1}{2}$$

Note: a **hypotenuse** can never be negative.

5.21. Exponential, Logarithmic, and Trigonometric Connections

---

The example below examines an exponential problem in tabular format. The standard equation for exponential equations is $y = a * b^x$.

Example 5.21.a:

Example 5.21.b: Determine the formula of the exponential function whose values are in the table below.

A. $f(x) = 5*3^x$

B. $f(x) = 1*(\frac{1}{5})^x$

C. $f(x) = 3*5^x$

D. $f(x) = 5*(\frac{1}{3})^x$

x	f(x)
-1	3/5
0	3
1	15

Example 5.21.c: The function $f(x) = \sin x$ is translated $\frac{\pi}{2}$ units to the right, 1 unit up, and its period is doubled. Which function below represents the new graph of $f(x)$?

A. $2\sin 4\pi\left(x - \frac{\pi}{2}\right) + 1$

B. $1\sin\frac{1}{2}(x+1) + \frac{\pi}{2}$

C. $1\sin\frac{1}{2}\left(x - \frac{\pi}{2}\right) + 1$

D. $\frac{\pi}{2}\sin 4\pi(x-1) + 1$

Solutions to Example Problems:

Solution to 5.21.a:

The answer is C. Working backwards is a viable method to use this problem if you forget how to solve problems such as this. To solve this problem, start by taking $y = a * b^x$ and plugging in the coordinate (0, 3) from the table above:

$$3 = a * b^0 \Rightarrow 3 = a$$

In taking advantage of the 0 value for x, we can determine "a" as shown in the line above. Next, pick any point and plug in a, x, and y in order to solve for b. (As a side note, the only answer choice with 3 in the "a" position is C so you could use process of elimination to cancel out the other answer choices).

$$15 = 3 * b^1 \Rightarrow \frac{15}{3} = b \Rightarrow 5 = b$$

Thus, $y = 3 * 5^x$

Solution to 5.21.b:

The answer is C. The standard formula for the sine functions is:

$$f(x) = \text{A} \sin B(x - C) + D$$

where A is the amplitude, $\frac{2\pi}{B}$ is the period, C is the horizontal shift (a negative C indicates a shift to the right and a positive C indicates a shift to the left) and D is the vertical shift.

Note that the coefficient before the x needs to be a 1 – thus you may have to factor a number out of $(x - C)$ in order to get to the correct standard form equation.

Also note that this formula works for cosine as well (just substitute in cos for sin).

First let's find B. The **period**, or time it takes the function to run through a single cycle, is 2π for a normal sine function. Thus, our doubled period is 4π. Plugging this into the function $\frac{2\pi}{B}$, we have:

$$\frac{2\pi}{B} = 4\pi \Rightarrow B = \frac{2\pi}{4\pi} \Rightarrow B = \frac{1}{2}$$

Thus, we plug in $\frac{1}{2}$ for B, $\frac{\pi}{2}$ for C, and 1 for D (and 1 for A since the amplitude is unaffected:

$$f(x) = 1 \sin \frac{1}{2}\left(x - \frac{\pi}{2}\right) + 1$$

5.22. **Solving Exponential, Logarithmic, and Trigonometric Equations**

The example below is a difficult question using the graph of an exponential equation of the form $y = a * b^x$. Plug in what you can and see if you can determine a formula. Hint: take advantage of the point lying on the y-axis first.

Example 5.22.a: What is the exponential form of the equation that represents the graph below?

A. $5 * (\sqrt[3]{3})^x$

B. $3 * (\sqrt[5]{1/3})^x$

C. $3 * (\sqrt[5]{3})^x$

D. $5 * (\sqrt[3]{1/3})^x$

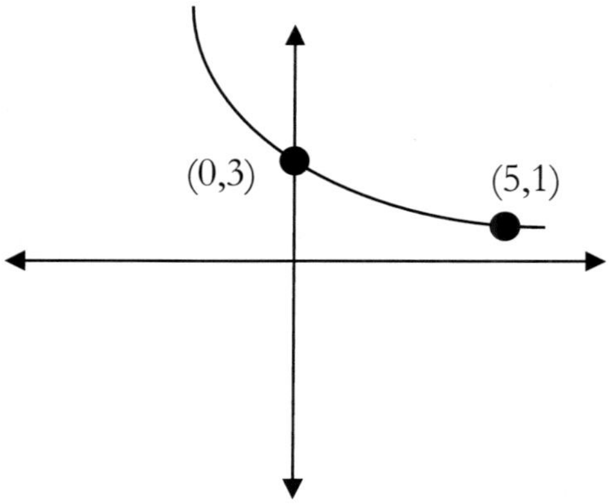

Solutions to Example Problems:

Solution to 5.22.a:

The answer is B. Please note that on the exam, you will not be given the standard form equation for an exponential function ($y = a * b^x$), and you may or may not be told that this is an exponential function.

In this problem, we have two coordinates to guide us: (0,3) and (5,1). Start, once again, by plugging in the coordinate with 0 as the x coordinate:

$$3 = a * b^0 \Rightarrow 3 = a$$

Next, we plug in the second coordinate and plug in 3 for a:

$$1 = 3 * b^5 \Rightarrow \frac{1}{3} = b^5 \Rightarrow \left(\frac{1}{3}\right)^{\frac{1}{5}} = b \Rightarrow \sqrt[5]{1/3} = b$$

Therefore, $y = 3 * \left(\sqrt[5]{\frac{1}{3}}\right)^x$.

5.23. Real-World Exponential, Logarithmic, and Trigonometric Functions

---

I chose the financial example below because of the emphasis that high school curricula has on teaching problems that use compounding interest. It is highly likely that the Praxis II ® (0061) exam will have one of these types of questions.

Financial problems are easiest solved using the **TVM solver** option on a Texas Instruments calculator. Here's a quick guide:

Go to APPS → Finance → TVM Solver

N = number of payments (if 3 years, N=3x12 months = 36)

I% = percent **interest**

PV = present value. (for loans you get, this number will be negative)

PMT = payment amount

FV = future value (for loans you get, this number will be 0)

P/Y = number of payments per year

C/Y = number of times compounded per year

PMT: End vs. Begin (makes payments at beginning or end of the month?) In general, use END

Leave the part of the TVM problem solver blank that you wish to solve for, and hit ALPHA→Enter on your Calculator to solve for it.

Example 5.23.a: Luke gets a $4000 loan from his Dad to pay for a pony which he has wanted forever. His Dad charges him 6% interest. Luke has to pay the loan back in 5 years, or else, and Luke must make a payment every month. How much will Luke's monthly payments be if he makes 1 payment each month and the interest is compounded monthly?

A. $68.15
B. $71.32
C. $66.67
D. $77.33

Solutions to Example Problems:

Solution to 5.23.a:

The answer is D. My TVM problem solver looks like this:

N = 60 (5 years x 12 months)

I% = 6

PV = -4000 (like negative 4000 dollars in Luke's pocket)

PMT = use the solve command here when you have everything else filled out (alpha+enter)

FV = 0

P/Y = 12

C/Y = 12

PMT: END

When I enter the solve command in the PMT area, I get 77.33.

Chapter 6

Measurement and Geometry

6.1. Error in Measurement

Mathematicians use precision and rounding error to describe the accuracy of measurements. The Praxis II ® (0061) exam may ask questions like: "how many **significant** digits are there in the measurement" or they may ask more difficult questions involving calculating percentage error or rounding error. For the problem below, use the formula:

$$percentage\, error = \frac{(actual\, value - approximated\, value)}{actual\, value} x100$$

Example 6.1.a: Luke rounds the number 100.4 to 100. What is the percentage error in this rounding?

A. .398%
B. .4%
C. .402%
D. 4%

Solutions to Example Problems:

Solution to 6.1.a:

The answer is A. Using the formula given, we have:

$$percentage\,error = \frac{(100.4-100)}{100.4} x100 = .398\% error$$

6.2. Converting Units

I convert units using **Stoichiometry** so that I can keep track of units effectively. To do this, use perfectly aligned fractions where units can be crossed out as you progress from one unit of measurement to another. The example below is a trigonometry problem where Stoichiometry will be needed to convert units.

Example 6.2.a: Luke is trying to find the height of the flagpole at school. He starts at the base of the pole, walks 30 lukefeet (the length of Luke's left foot) and determines that the top of the pole is 60 degrees up from his eye level. If 1 lukefoot = 1.2 actual feet and Luke's eye level is 6 lukefeet high, how tall is the flagpole?

A. 52 feet
B. 58 feet
C. 64 feet
D. 70 feet

Solutions to Example Problems:

Solution to 6.2.a:

The answer is D. I'm going to make a sketch for this to make sure my trigonometry is accurate:

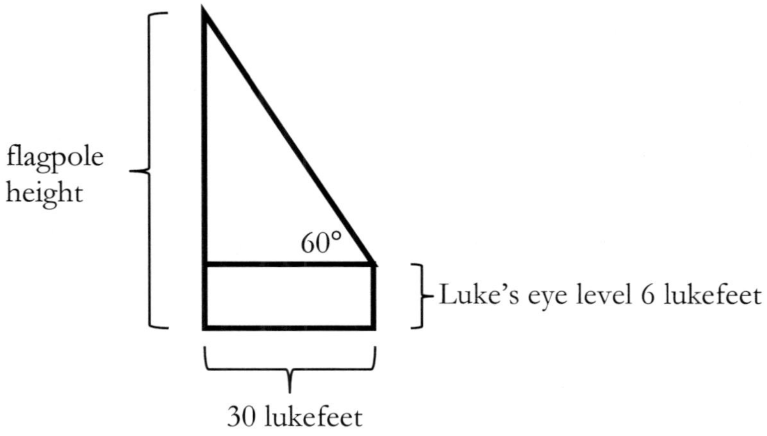

If the 60 degree angle is theta, then the flagpole height minus Luke's eye level is the opposite side and the distance that Luke walked is the adjacent side. Using tangent, I get:

$$tan\theta = \frac{opposite}{adjacent} \Rightarrow \tan(60) = \frac{flagpole-6}{30} \Rightarrow .1.73(30) = flagpole-6$$

Thus, $51.96+6 = flagpole\,height = 57.96$ is the height of the flagpole in lukefeet.

However, this question asks for an answer in feet rather than in lukefeet. Set up the conversion using stoichiometry so that the units on top and on bottom cancel each other out:

$$57.96\,lukefeet\;x\frac{(1.2\,feet)}{1\,lukefoot} = 69.55\,feet$$

6.3. Similarity and Triangulation

Similarity is a useful tool in decoding difficult geometry problems; it often involves setting up algebraic proportions and solving those proportions. As a layman definition, similar triangles are those that are of the same proportions. If a triangle can be shrunk or enlarged to fit perfectly on top of another triangle without compromising its angles, then the triangles are similar.

Triangles are similar if they have two equivalent angles (AA). Also, triangles which are congruent are similar.

Example 6.3.a: Find the value of x in the figure below. Given: line segment $\overline{AF}$ is parallel to line segment $\overline{EG}$, line $\overline{AF}$ = 4, line $\overline{EG} = x - 2$, line $\overline{CA} = 6$, line $\overline{EC} = 9$.

A. 7
B. 8
C. 4
D. 6

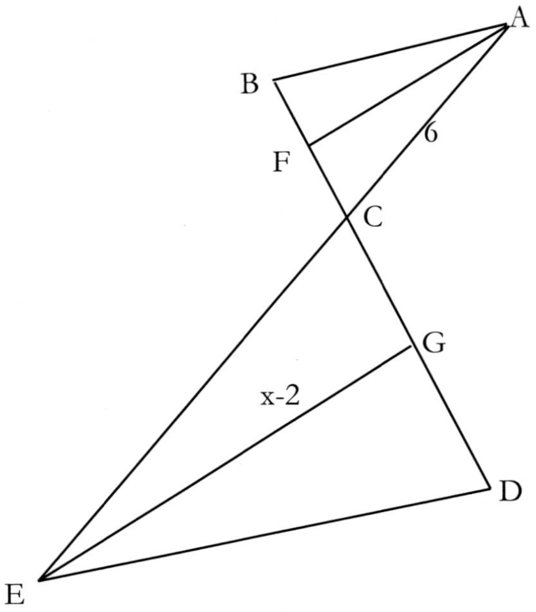

Solutions to Example Problems:

Solution to 6.3.a:

The answer is B. As you might guess from reading the clues above, the triangles EGC and AFC are similar.

You can prove that triangles EGC and AFC are similar using the **Angle-Angle** theorem: First, since angle FCA and angle GCE are vertical angles, they are congruent. Secondly, since line segments $\overline{AF} \text{ and } \overline{EG}$ are parallel, angles FAC and GEC are congruent because they are **alternate interior angles** of **parallel** lines. Thus, we have shown that, through Angle-Angle, triangles ECG and ACF are similar.

Because triangles ECG and ACF are similar, we can compare **corresponding parts** through using proportions. We have that:

$$\frac{\overline{EG}}{\overline{AF}} = \frac{\overline{EC}}{\overline{AC}} \Rightarrow \frac{x-2}{4} = \frac{9}{6}$$

Cross multiplying, we get $(9)(4) = 6(x-2) \Rightarrow 36 = 6x - 12 \Rightarrow 48 = 6x \Rightarrow 8 = x$.

6.4. Points, Lines, and Planes

You need mostly common sense when it comes to points, lines, and planes. However, knowing that $\text{the sum of the interior angles of a polygon} = (n-2)180\,degrees$ can be quite useful.

Example 6.4.a: What is the measurement of each exterior angle of a convex equiangular pentacontagon (a 50 sided polygon)?

A. $172.8\,degrees$

B. $187.2\,degrees$

C. $8640\,degrees$

D. $191\,degrees$

Solutions to Example Problems:

Solution to 6.4.a:

The answer is B. Using the formula above, we have that the sum of all the INTERIOR angles of the pentacontagon is:

$$sum\,of\ interior\ angles = 180(50-2)\,or\,180(48)\,or\,8640\,degrees$$

Therefore, each of the 50 angles inside has a measurement of $\frac{8640\,degrees}{50\,angles} = 172.8\,degrees\ per\ interior\ angle$. Each interior angle has this measurement because the **polygon** is equiangular meaning that each interior angle is identical.

If each interior angle is 172.8 degrees, then each **exterior angle** has a measurement of $360-172.8\,degrees = 187.2\,degrees$.

6.5. Pythagorean Theorem

The Pythagorean theorem could possibly be the most important concept to know for all of geometry on the Praxis II ® (0061) exam. $a^2+b^2=c^2$ where a and b are side lengths and c is the hypotenuse of a right triangle. Use the Pythagorean theorem to find the distance between any two points on a **Cartesian** coordinate plane rather than memorizing the distance formula.

The problem below represents the trickery that you may experience on the Praxis II ® (0061) exam.

Example 6.5.a: Luke wishes to chop the top off of an 8-ball and then mount that 8-ball on the front of his bicycle (he's cool like that). He saws the top off of an 8-ball such that a circle is made by the saw (represented by a plane below); the circle formed has a diameter of 10. If the diameter of the 8-ball is 16, what is the value of x which represents the distance from the center of the 8-ball to the center of the circle created by Luke's saw-cut?

A. $3\sqrt{5}$
B. 5
C. 8
D. $\sqrt{39}$

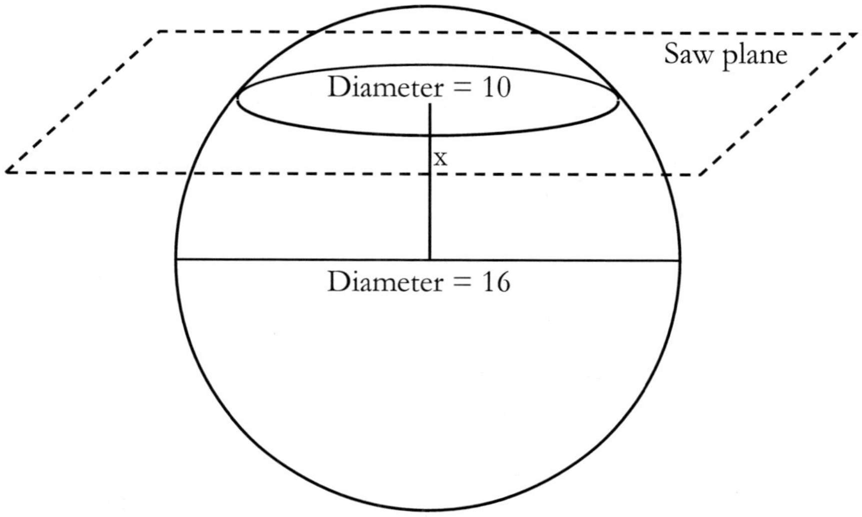

Solutions to Example Problems:

Solution to 6.5.a:

The answer is D. This problem looks particularly nasty until you realize that the distance between the center of the 8-ball and any point on the outside of the 8-ball is 8. The diagram below trims the fat from this problem:

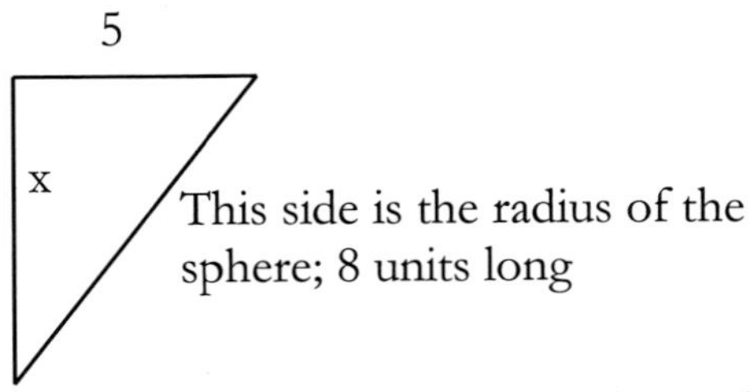

Thus, we can simply use the Pythagorean theorem to solve the problem:

$$x^2 + 5^2 = 8^2 \; or \; x^2 + 25 = 64 \; or \; x^2 = 39 \; or \; x = \sqrt{39}$$

6.6. Similarity and Congruence

Just as triangles can be similar, so can all polygons.

For a polygon to be congruent, it must be the same size AND shape as another. The example below tests your knowledge of the law of reflection. Hint: try to use similar triangles.

Example 6.6.a: Doni is trying to set up a hidden camera so that he can bust Luke when Luke steals Doni's stash of candy. To do this, Doni sets up a long mirror in the hallway between his bedroom and the living room as shown below; the mirror lies perpendicular to both the hidden camera and to the candy stash. Where must Doni aim his hidden camera in order to have it point directly at his candy stash?

A. $\frac{75}{8}$ feet from the left edge of the mirror

B. 8 feet from the left edge of the mirror

C. $\frac{26}{3}$ feet from the left edge of the mirror

D. $\frac{23}{3}$ feet from the left edge of the mirror

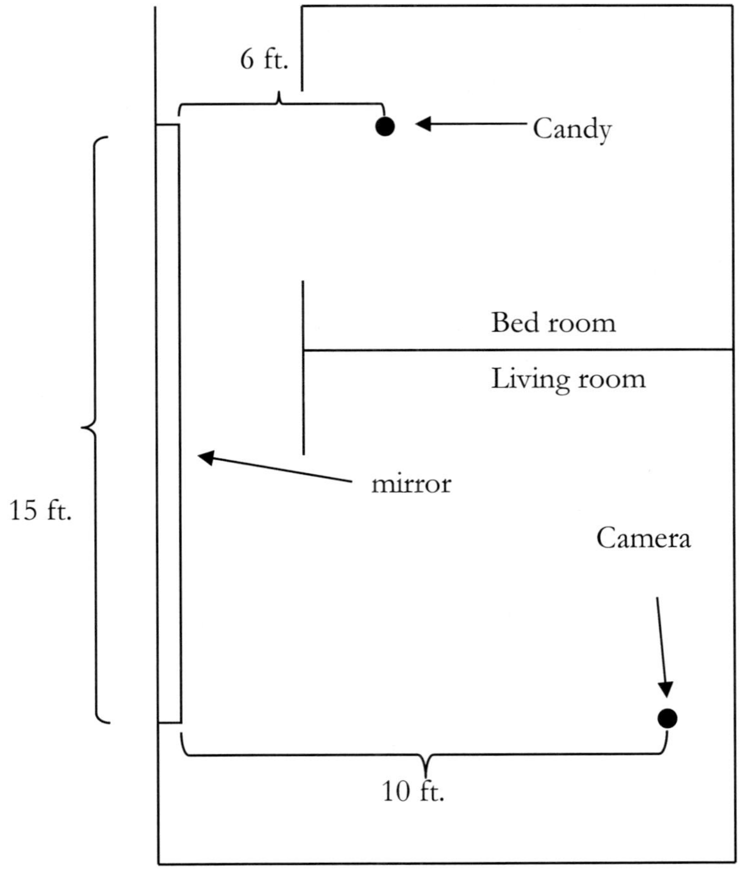

Solutions to Example Problems:

Solution to 6.6.a:

The answer is A. What we aim to do here is create two perfectly similar triangles. I will draw them below and label a few points to make the answer description easier:

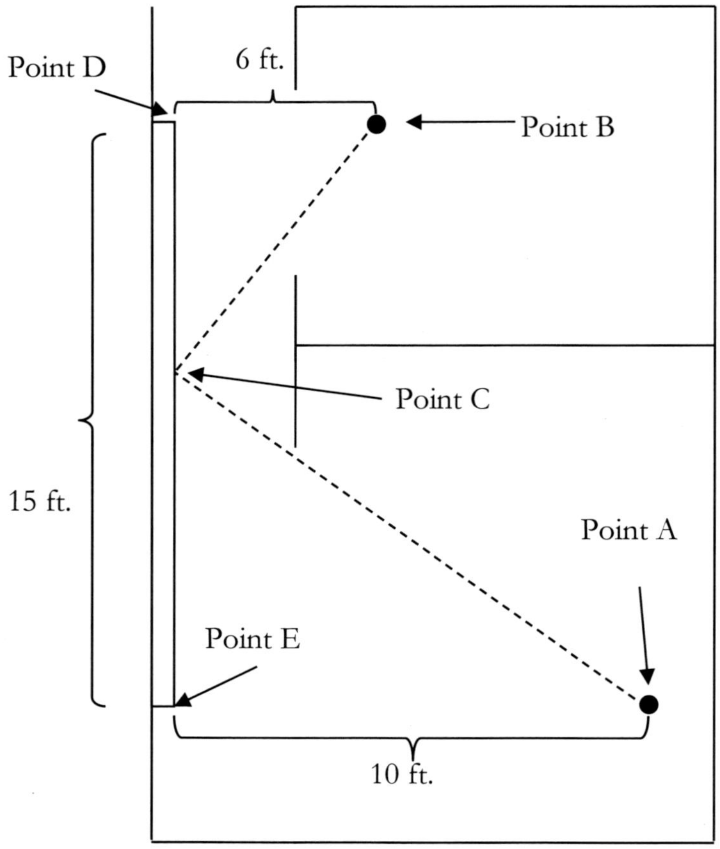

First, note that $\overline{DC} = 15 - \overline{EC}$. Triangle DBC must be similar to triangle EAC. Thus, we can set up a proportion in order to find the length $\overline{EC}$ which will be our answer:

$$\frac{6}{\overline{DC}} = \frac{10}{\overline{EC}}$$

Substituting $15 - \overline{EC}$ in for $\overline{DC}$, we get:

$$\frac{6}{15 - \overline{EC}} = \frac{10}{\overline{EC}}$$

Simplifying further, we get:

$$6\overline{EC} = 10\left(15 - \overline{EC}\right) \Rightarrow 6\overline{EC} = 150 - 10\overline{EC} \Rightarrow 16\overline{EC} = 150$$

$$\text{Thus, } \overline{EC} = \frac{75}{8} \textit{ feet}$$

6.7. Nets, Projections, and Cross Sections

Nets are two-dimensional figures that can be folded up to form polyhedrons (which are three-dimensional). Exam writers can create problems that twist, turn, reflect, and translate two and three-dimensional figures back and forth ask difficult questions about them. Try this one.

Example 6.7.a: Name the solid that can be folded from the net below.

A. hexagon
B. kite
C. hexagonal prism
D. hexagonal cylinder

Solutions to Example Problems:

Solution to 6.7.a:

The answer is C. You can use process of elimination to cancel out answers A and B even if you don't know that this three dimensional figure is a hexagonal prism.

6.8. Triangles and Quadrilaterals

A **median** of a triangle connects the vertex of a triangle to the midpoint of the opposite side.

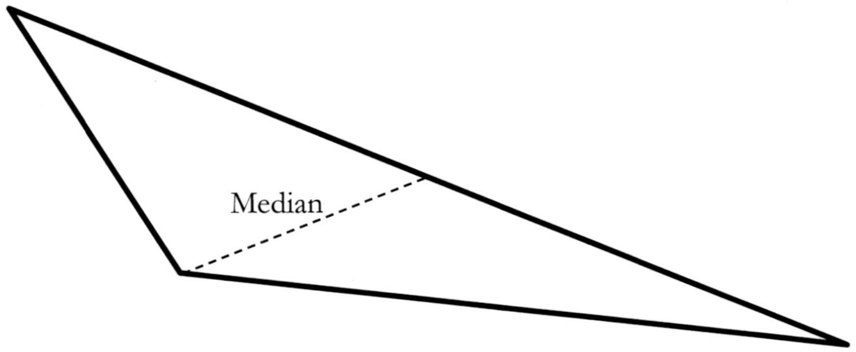

A **midpoint of a triangle or a quadrilateral** is the point on the side of a triangle that exactly in the middle between two vertices.

An **altitude** of a triangle connects the vertex of a triangle to the opposite side at a 90 degree angle. The altitude of a triangle can be on the outside of a triangle as in the diagram below. The point where all three altitudes meet is called the **orthocenter**.

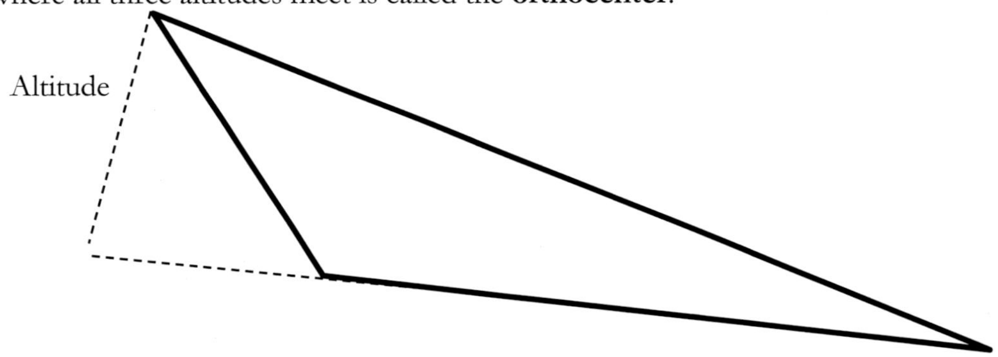

The following section discusses the properties of quadrilaterals, but here is an interesting question to peak your interest in special quadrilaterals!

Example 6.8.a: Given a quadrilateral, ABCD, with midpoints, WXYZ, where W is the midpoint of $\overline{AB}$, X is the midpoint of $\overline{BC}$, Y is the midpoint of $\overline{CD}$, and Z is the midpoint of $\overline{DA}$, which is true?

A. $\overline{ZW}$ and $\overline{WX}$ are perpendicular.
B. $\overline{ZW}$ and $\overline{YX}$ are parallel.
C. $\overline{ZW}$ and $\overline{YX}$ are perpendicular.
D. There is not enough information to prove any of the above.

Solutions to Example Problems:

Solution to 6.8.a:

The answer is B. First, draw a non-regular quadrilateral and draw in its midpoints with all the connecting lines:

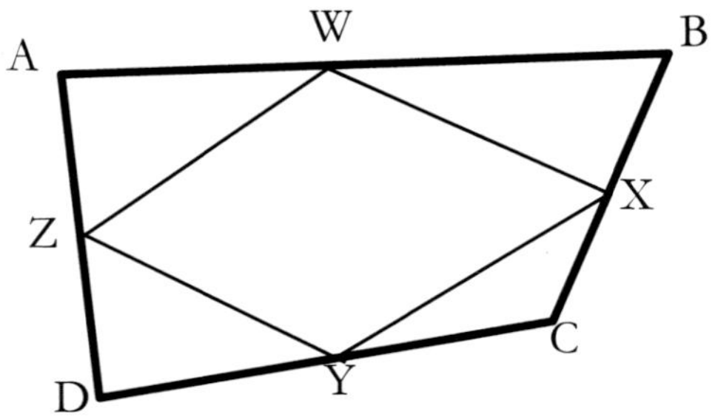

This trick to proving this dilemma is to realize that $\overline{ZW}$ is the midsegment of triangle DAB which means that $\overline{ZW}$ is parallel to $\overline{DB}$ (by the midsegment theorem on triangles). Also, $\overline{XY}$ is the midsegment of triangle DBC which means that $\overline{XY}$ is also parallel to $\overline{DB}$. Since to $\overline{ZW}$ and $\overline{YX}$ are parallel to the same line, they must also be parallel to each other.

6.9. Relationships Among Sets of Special Quadrilaterals

The Venn diagram below is a visual synopsis of all sets of special quadrilaterals:

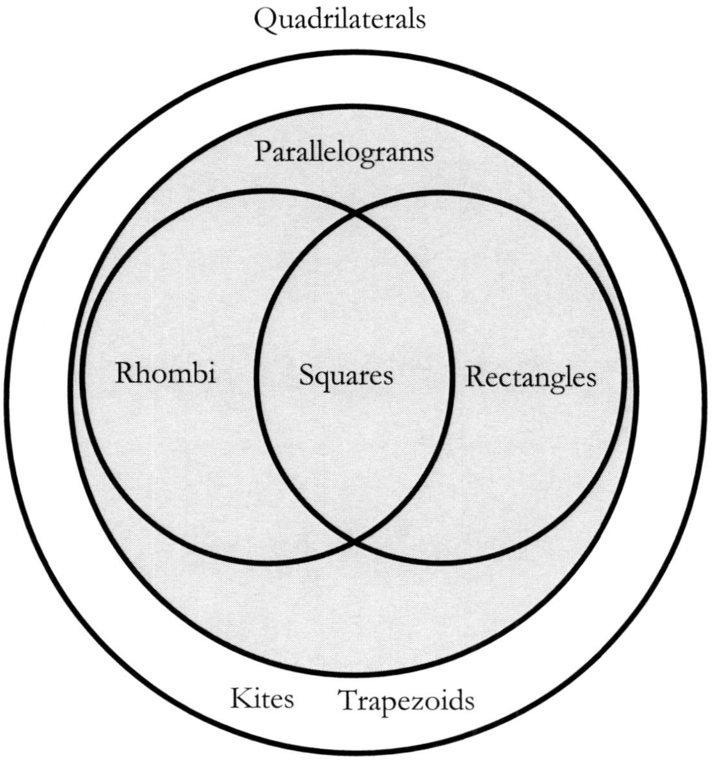

Basically, here is a rundown: everything in the diagram above is a quadrilateral.

Rhombi, **squares**, and **rectangle**s are parallelograms meaning that they have two sets of opposite sides that run parallel to each other.

Rhombi are quadrilaterals that have all four sides being equal so all squares are rhombi, but not all rhombi are squares.

Rectangles have four sides, all meeting at 90 degree angles so all squares are also rectangles but not all rectangles are squares.

For the problems below: $\subseteq$ means "subset", and $\in$ means "belongs to." Also note that for two sets to be equal "=" the two sets must be subsets of each other.

Example 6.9.a: Let Q be the set of all quadrilaterals, P be the set of all parallelograms, S be the set of all squares, R be the set of all rhombi, T be the set of all trapezoids, K be the set of all kites, and E be the set of all rectangles. Then which of the following is true?

A. $P = \{R, S\}$
B. $R, S \in P$
C. $P \subseteq T$
D. $R = E$

Example 6.9.b: Let Q be the set of all quadrilaterals, P be the set of all parallelograms, S be the set of all squares, R be the set of all rhombi, T be the set of all trapezoids, K be the set of all kites, and E be the set of all rectangles. Then which of the following is **not** true?

A. $P = \{R, E\}$

B. $R, E, S, T, K \in Q$

C. $Q = \{P, K, T\}$

D. $S = R, E$

Solutions to Example Problems:

Solution to 6.9.a:

The answer is B. I busted out a little set theory math for this problem because it is important that you learn some of the set theory language. $R, S \in P$ reads "R and S belong to P" which is true - the set of rhombi and squares do belong to the set of parallelograms.

The trick answer in this problem was answer A. To say that $P = \{R, S\}$ is to say that the set of parallelograms is equal to the set of rhombi and squares; this is not true because P also encompasses the set of all rectangles. In other words, even though {R, S} is a subset of P, P is not a subset of {R, S} and therefore the two sets cannot be equal.

The little symbol $\subseteq$ in answer C means "subset" and P is not a subset of T.

In set language, for two sets to be equal, like in answer D, the two sets would need to be subsets of each other. In other words, if $R \subseteq E$ and if $E \subseteq R$, then $R = E$. This, of course, is not true.

Solution to 6.9.b:

The answer is D. The set of all squares does NOT equal rectangles and rhombi.

Answer A is correct although it may not look right. Since R and E encompass all squares, R plus E is equal to P.

Answer B is correct. All those sets belong to the set of Quadrilaterals.

Answer C is correct for similar reasons that answer A is correct: P,K, T encompasses all of set Q.

6.10. Properties of Circles

Inscribed angles are angles in which the vertex lies on the circumference of the circle like $< ABC$ in the following diagram. The sides of the inscribed angle are **chords**; $\overline{AB}$ and $\overline{BC}$ are chords (the straight lines). AB is an **arc** the follows the circumference of the circle.

In most cases, the **intercepted arc** created by the inscribed angle is twice the degree measurement as the angle itself: $AC = 2$ times the measure of $< ABC$; the only case where this is not true is when the vertex of the inscribed angle is inside the arc itself.

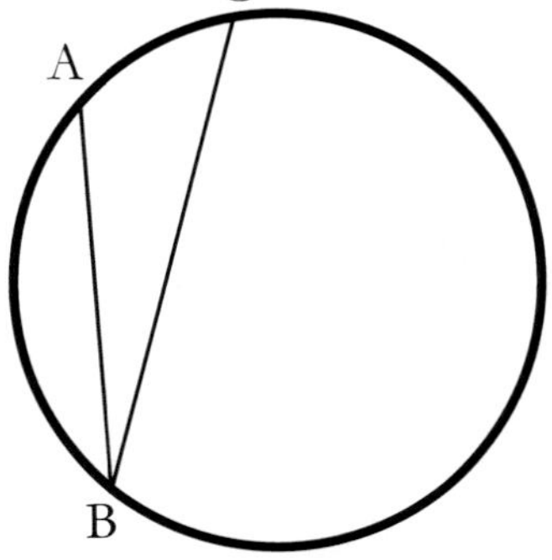

A **central angle**'s vertex is at the center of a circle.

A **tangent line** is a straight line that touches the outside of the circle exactly once; this means that a diameter line of a circle which intersects a tangent line must intersect at a right angle to the diameter line. Also note that if two tangent lines intersect, as in the following figure, then the distances from the intersection of the tangent lines to the circle must be congruent: $\overline{AC} = \overline{AB}$.

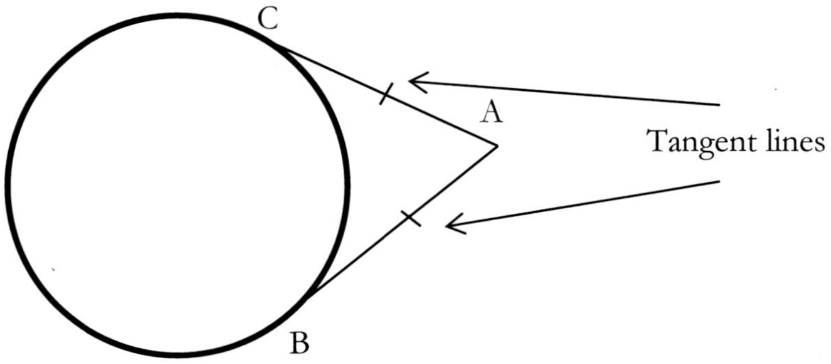

Sectors and segments of a circle are shown in the following diagram where the vertex of each lies at the center of each circle:

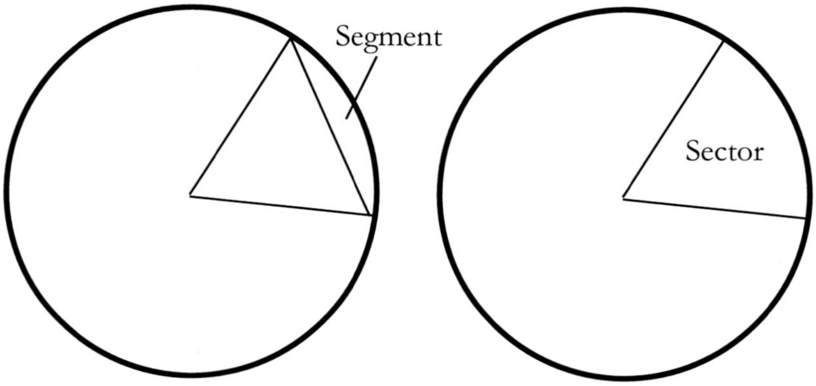

Example 6.10.a: What is the perimeter of the triangle below given that all the sides of the triangle are tangent to the circle and each measurement given below represents the distance from the vertex of the triangle to the intersection between the triangle and the circle? Note: not drawn to scale.

A. 34
B. 36
C. 38
D. Not enough information given.

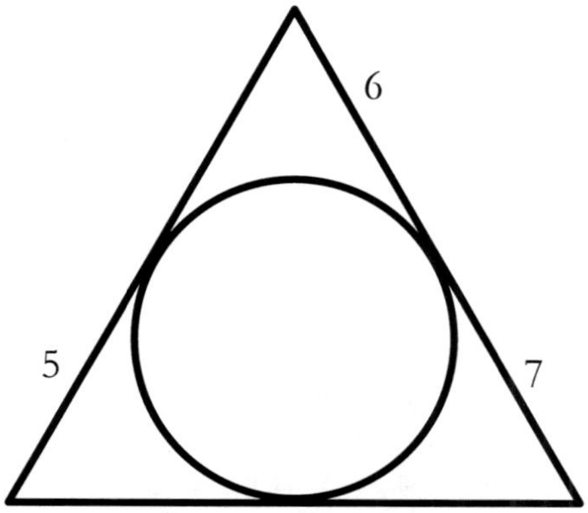

Example 6.10.b: Find the area of a sector made by a central angle of 40 degrees in a circle with radius, r.

A. $\frac{\pi r^2}{9}$
B. $40\pi r^2$
C. $\frac{\pi r^2}{8}$
D. $\frac{\pi r^2}{40}$

Solutions to Example Problems:

Solution to 6.10.a:

The answer is B. Since the tangent line from each vertex must be equidistant, we can re-label:

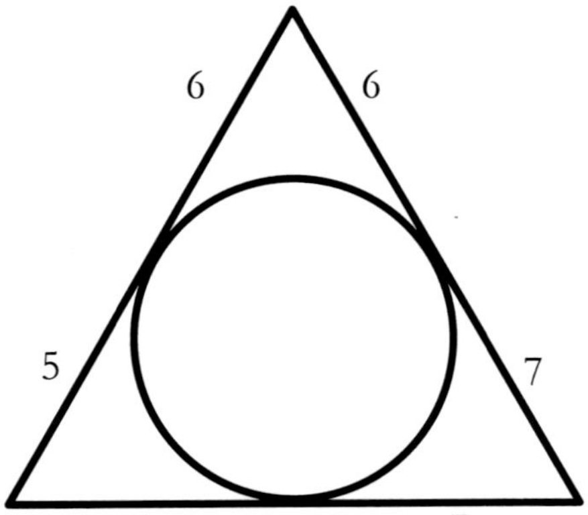

Thus, the perimeter is $5+5+6+6+7+7=36$.

Solution to 6.10.b:

The answer is A. The area of the entire circle would be $A=\pi r^2$ but we are only looking at a sector whose central angle is 40 degrees rather than 360 degrees. We know that 40 degrees is 1/9 the measurement of 360 degrees. Therefore the sector area will be $\frac{\pi r^2}{9}$.

6.11. Translations, Rotations, and Reflections

A translation is a vertical or horizontal movement in a two-dimensional plane. A rotation is a movement around an axis point and a reflection is a perfect flip over an axis line.

Example 6.11.a: A translation of $\overline{AB}$ onto $\overline{AB}'$ (read as "line segment AB prime") is described by vector $\vec{XY}$. Find the value of each variable if:

$\vec{XY}$ = <2,3>; $A=(-2,r)$; $A'=(4s-2,3)$; $B=(2q+1,1)$; $B'=(1,4t)$

A. $t>r>s>q$
B. $r>t>q>s$
C. $t=q>r>s$
D. $t>s>r>q$

Example 6.11.b: The cube shown below is plotted on an x, y, z coordinate system and has side lengths of 2. If this cube is rotated 180 degrees about the y-axis followed by a rotation of 180 degrees about the z-axis, where will point A be?

A. $(2,2,-2)$
B. $(-2,-2,-2)$
C. $(2,2,2)$
D. $(-2,2,-2)$

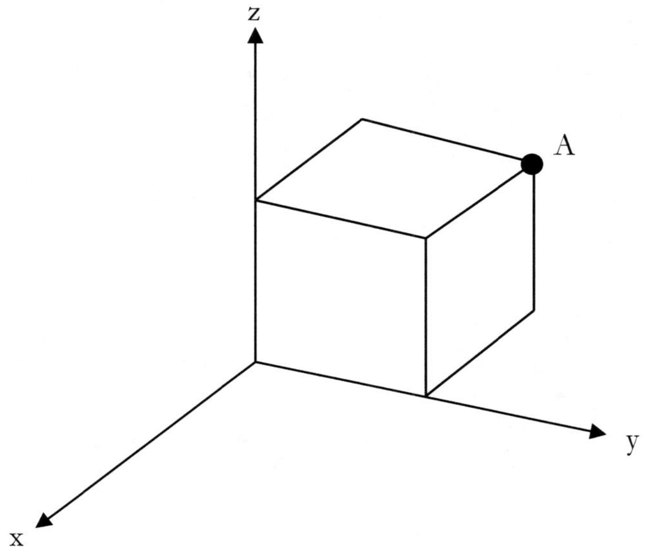

Solutions to Example Problems:

Solution to 6.11.a:

The answer is D. Let's start with point A which begins at point $(-2,r)$, and is moved 2 right and 3 up via vector XY, and ends up at $(4s-2,3)$. We can write a system of equations from this:

$$-2+2=4s-2$$

$$r+3=3$$

Solving for our first two variables, we have that:

$$0=4s-2 \Rightarrow 2=4s \Rightarrow \frac{1}{2}=s$$

$$r=0$$

Let's set up a system of equations for point B:

$$2q+1+2=1$$

$$1+3=4t$$

Solving, we get:

$$2q+3=1 \Rightarrow 2q=-2 \Rightarrow q=-1$$

$$4=4t \Rightarrow 1=t$$

Therefore, $s=\frac{1}{2}, r=0, q=-1, t=1 \Rightarrow t>s>r>q$

Solution to 6.11.b:

The answer is B. This problem can be particularly tricky even if you try to sketch out the cube as it rotates about the 2 axes. Point A is currently at $(-2,2,2)$ where (x,y,z). Rotating the cube about the y-axis puts point A at $(2,2,-2)$ because it changes the sign of the two axes which are not being rotated around – in this rotation, the x-axis and the z-axis. A rotation about the z-axis would then put point A at (-2, -2, -2).

6.12. Slope, Distance, and Midpoint

Slope is rise over run. Slope can be found using the formula: $m=\frac{dx}{dy}=\frac{rise}{run}=\frac{y_2-y_1}{x_2-x_1}$

The distance formula is: $d=\sqrt{(x_2-x_1)^2+(y_2-y_1)^2}$ but the distance between any two points should be determined with the Pythagorean Theorem; to do this, make a right triangle out of the two points and use $a^2 + b^2 = c^2$. When you solve for c in the Pythagorean Theorem, you derive the distance formula as shown above. I use the Pythagorean Theorem to find distance because it saves me from memorizing a formula.

The midpoint formula is simply:

$(\textit{the average of the } x\textit{-coordinates}, \textit{the average of the } y\textit{-coordinates})$ or $(\frac{x_1 + x_2}{2}, \frac{y_1 + y_2}{2})$

I included the second problem below because there is a small chance that the test writers will include a three-dimensional distance problem.

Example 6.12.a: Line segment $\overline{AB}$ is defined as $(0,1)\,(3,4)$ and line segment $\overline{CD}$ is defined as $(-2,-3)(-4,1)$. What is the distance between the midpoints of line segment $\overline{AB}$ and line segment $\overline{CD}$?

A. $\frac{\sqrt{130}}{2}$

B. 6

C. $\sqrt{41}$

D. $3\sqrt{5}$

Example 6.12.b: What is the distance from the corner of a 3x4x5 inch box to its center?

A. 3 in.

B. $\sqrt{50}$ in.

C. $\frac{5}{2}$ in.

D. $\frac{5}{2}\sqrt{2}$ in.

Solutions to Example Problems:

Solution to 6.12.a:

The answer is A. First, let's find the two midpoints. The midpoint of line $\overline{AB}$ is $\left(\frac{0+3}{2},\frac{1+4}{2}\right) or \left(\frac{3}{2},\frac{5}{2}\right)$. The midpoint of line $\overline{CD}$ is $\left(\frac{-2-4}{2},\frac{-3+1}{2}\right) or\ (-3,-1)$. Then we take the distance between those two midpoints:

$$distance = \sqrt{\left(\frac{3}{2}-(-3)\right)^2+\left(\frac{5}{2}-(-1)\right)^2}\ or\ \sqrt{\left(\frac{9}{2}\right)^2+\left(\frac{7}{2}\right)^2}\ or\ \sqrt{\frac{81}{4}+\frac{49}{4}}\ or\ \sqrt{\frac{130}{4}}$$

Reducing this, we get $\frac{\sqrt{130}}{2}$.

Solution to 6.12.b:

The answer is D. The distance formula for finding distance between two points in **three dimensional** space is $d=\sqrt{(x_1-x_2)^2+(y_1-y_2)^2+(z_1-z_2)^2}$. In the example above, we can substitute 3 for x_1-x_2, 4 for y_1-y_2, and 5 for z_1-z_2 as these are the lengths of those sides. Thus, we have $d=\sqrt{3^2+4^2+5^2}\ or\ d=\sqrt{50}$. Of course, this represents the distance from one corner to the other; we must halve our answer to find the distance between a corner and the center of the box: $d=\frac{1}{2}\sqrt{50}\ or\ d=\frac{1}{2}(5)\sqrt{2}\ or\ d=\frac{5}{2}\sqrt{2}$.

6.13. Connecting Algebra and Geometry

Algebra and geometry are easily tied together. The example below is a problem you may find as an introductory problem at the beginning of a calculus chapter on integrals. If you square both sides and group the x's and y's, this becomes the standard form for a geometric figure. Can you tell which one?

Example 6.13.a: What is the area between the curve $y=\sqrt{16-x^2}$ and the x-axis?

A. 4π
B. 6π
C. $8\pi/3$
D. 8π

Solutions to Example Problems:

Solution to 6.13.a:

The answer is D. If you change the form of $y=\sqrt{16-x^2}$ *into* $y^2=16-x^2 \Rightarrow y^2+x^2=16$, you may be able to tell that the curve makes a **semi-circle** with a radius of 4. Thus, the area is:

$$Area=\frac{1}{2}\pi r^2 \Rightarrow \frac{1}{2}\pi\left(4^2\right) \Rightarrow \frac{1}{2}\pi(16) \Rightarrow 8\pi$$

6.14. Axiomatic Systems

An axiomatic system is the basis for formal proofs in geometry; it is based upon the idea that each axiom (or self-evident truth) works in conjunction to prove a large idea.

Example 6.14.a: An axiomatic system is consistent if it:

A. lacks contradiction
B. is **infallible**
C. is **independent**
D. is complete

Solutions to Example Problems:

Solution to 6.14.a:

The answer is A. An axiomatic system is independent if each of its axioms is independent – that is, no axiom can be logically derived from another axiom in the system.

6.15. Deductive and Inductive Reasoning

Inductive reasoning allows for a false conclusion even if its premises are true while deductive reasoning attempts to draw conclusions that must be true based on premises than can be false. The easiest example of deductive reasoning I can think of is a Sudoku puzzle where each number in the puzzle is deduced by knowing that no other numbers can fit in a specific place.

In reasoning, a conclusion is valid even if its premises are false whereas a conclusion is sound if its premises are **sound** and it is derived from deductive reasoning.

Example 6.15.a: Luke knows that three high-fives are better than two high-fives. He also knows that four high-fives are better than three high-fives. Therefore, four high-fives must be better than two high-fives. This is an example of:

A. Common sense
B. Inductive reasoning
C. Deductive reasoning
D. Conjecture

Solutions to Example Problems:

Solution to 6.15.a:

The answer is C. The logic used here is deductive. An example of inductive reasoning might be: Luke has only received high-fives therefore all fives must be high-fives (clearly Luke is wrong – the low-five and the *Top Gun*-five demonstrate this).

6.16. Making Geometric Projections and Proving Conjectures

In mathematics, a conjecture is a statement which has not been proven correct, but appears to be so. Making, testing, justifying, and proving conjectures requires knowledge of deductive reasoning and the ability to recognize premises as true or false. Example conjecture: if Luke's hair looks shorter today, Luke must have gotten a haircut. You only need a single counter example to prove a conjecture false: Luke actually got his ears lowered.

Predictions and generalizations can also be make from patterns; be careful when making them not to make assumptions.

Example 6.16.a: In the figure below, what does a+b+c+d =?

A. Not enough information
B. 260 degrees
C. 280 degrees
D. 300 degrees

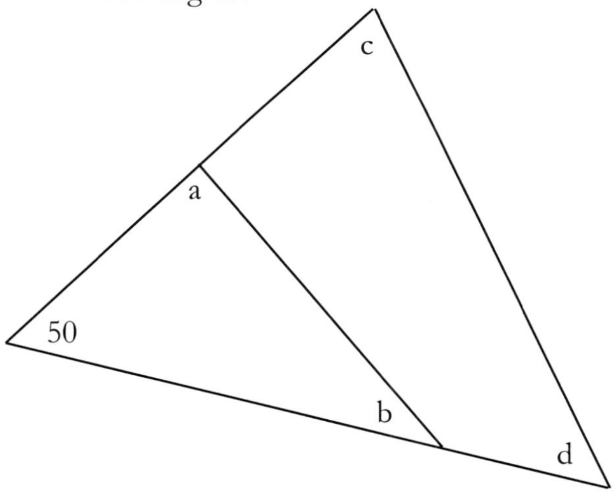

Example 6.16.b: Which is a counter example of the following conjecture: 6+7=13?

A. $6+7 \neq 13$
B. $13 = 20-7$
C. On a clock, $6+7=1$
D. $6+7=14$

Solutions to Example Problems:

Solution to 6.16.a:
The answer is B.

$$50 + measure < a + measure < b = 180; also, 50 + measure < c + measure < d = 180$$

Adding these together:

$$100 + a + b + c + d = 360 \Rightarrow a + b + c + d = 260 degrees.$$

Solution to 6.16.b:
The answer is C. The fact that $6 + 7 = 1$ on a clock is a perfect counterexample for the conjecture that 6+7 always equals 13.

6.17. Non-Euclidean Geometries

Non-Euclidean geometries are those geometries where Euclid's parallel postulate is not true; that is, in non-Euclidean geometries, given a line and a point not on the line, there will be either MORE than one parallel line which runs through the point or there will be NO lines that are parallel to the line and run through the point:

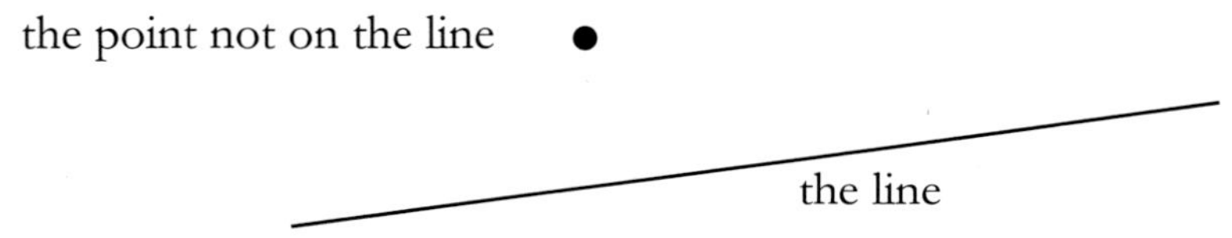

Note: the diagram above is equivalent to Euclid's parallel postulate.

Example 6.17.a: Which of the following postulates are examples of non-Euclidean geometries?

I. "Within a two-dimensional plane, for any given line ℓ and a point A, which is not on ℓ, there is exactly one line through A that does not intersect ℓ."
II. "On the surface of a sphere, the sum of the angles of a triangle is not equal to 180 degrees."
III. "Space is not flat, but elliptically curved near regions where energy is present."

A. I and III
B. I, II, and III
C. II and III
D. III only

Solutions to Example Problems:

Solution to 6.17.a:

The answer is C. Roman numeral I is equivalent to Euclid's parallel postulate and therefore a Euclidean geometry. Roman numerals II and III are only possible in geometries where Euclid's parallel postulate is not true.

Chapter 7

Probability and Statistics

7.1. Organizing data in ways appropriate to solving problems

You should know how to use and create the following types of plots and graphs: histograms, box and whisker plots, bar graphs, stem and leaf plots, scatter plots, and circle charts. You will see a discussion of each of these types of plots and graphs throughout this chapter, but for now the topic is histograms.

A **histogram** looks a lot like a bar graph except that a histogram has intervals on its x-axis rather than categories. Also, a histogram's intervals will be continuous. A histogram's y-axis will be a count of how many units are in each interval.

A note on histograms: numbers that lie between intervals are included in the HIGHER interval. For example, if a histogram has intervals of 10 then the actual intervals are: 0-9, 10-19, 20-29, etc. so the number 10 will lie between 10 and 20. The number 20 will lie between 20 and 30.

Here is an example histogram with intervals of 10. Notice that 30 goes between 30 and 40. The data set is ages of people in a train car: {2,3,5,10,12,15, 18, 22, 25,30}

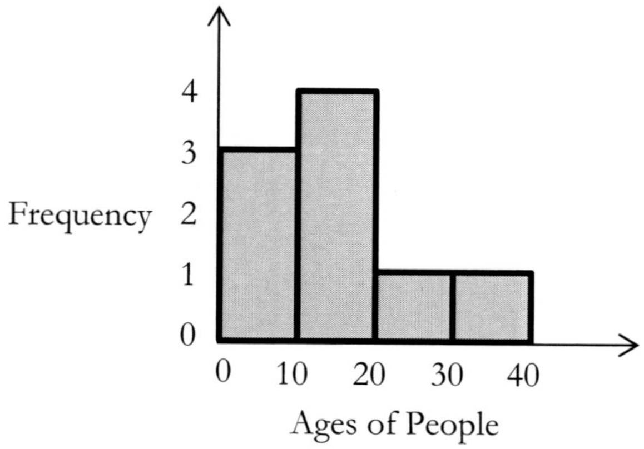

In slight contrast, you can also have a **frequency histogram**, which measure the percentage of units in each interval rather than the raw number. Therefore, the y-axis will be percentages rather than a count of how many units are in each category.

All that said on histograms, the question below concerns frequency bar graphs.

Example 7.1.a: According to the frequency bar graph below, what percent of people have a favorite color of red or blue?

A. 25%
B. 30%
C. 35%
D. 40%

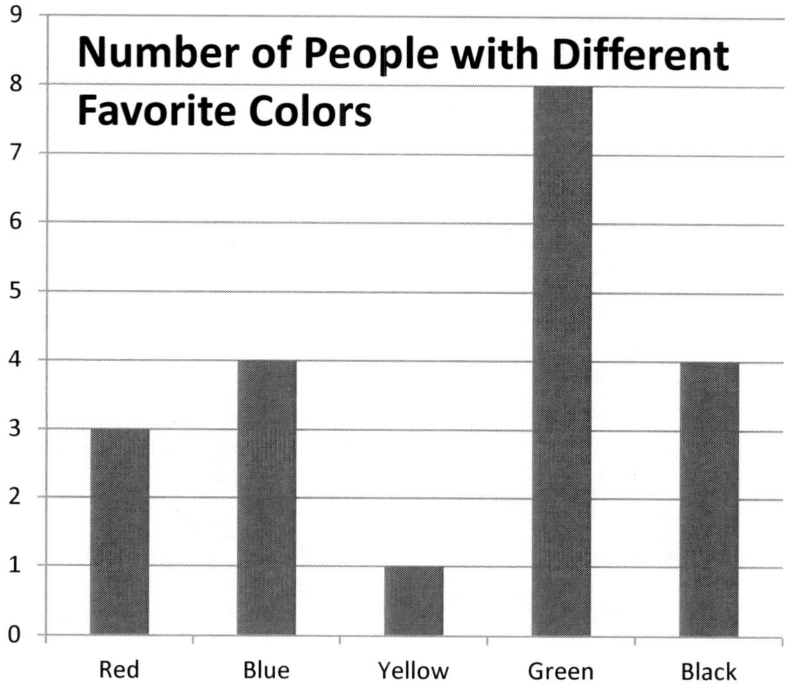

Solutions to Example Problems:

Solution to 7.1.a:

The answer is C. Here are a total of 7 people who have a favorite color of either red or blue out of a total of 20 people in the entire graph. Thus, the percentage of people whose favorite color is either red or blue is $\frac{7}{20}(100) = 35\%$.

7.2. Empirical Probabilities

Empirical probability is an estimate of the probability that an event will occur based on observations or empirical data.

Given that each outcome of an event is equally likely, statisticians use the following formula to determine the probability of an event:

$$P(E) = \frac{\textit{number of favorable outcomes}}{\textit{total number of possible outcomes}}$$

where P(E) represents the probability of an event, E. If two events are mutually exclusive, like in the example below, then we can add the probabilities of the two events together.

Example 7.2.a: If Luke draws a card from a standard 52 card deck, what is the probability that he will draw a queen OR a spade (drawing the queen of spades does NOT count)?

A. $\frac{15}{52}$

B. $\frac{17}{52}$

C. $\frac{4}{13}$

D. $\frac{19}{52}$

Solutions to Example Problems:

Solution to 7.2.a:

The answer is A. The probability of drawing a queen is $\frac{3}{52}$ (remember: the queen of spades doesn't count); the probability of drawing a spade is $\frac{12}{52}$ (again, no queen). Therefore, the probability that Luke will draw a spade or a queen is:

$$P = \frac{3}{52} + \frac{12}{52} = \frac{15}{52}$$

7.3. Principles And Techniques of Theoretical Probability

If an event is reliant upon the outcome of a previous event, then it is a **dependent event** (like choosing a card in a deck of cards where someone else has already chosen a card). If an event is not reliant upon the outcome of a precious event, it is an independent event (like rolling a die: previous rolls will not effect the odds in later rolls). When determining the likelihood of successive events, you multiply the probability of the first with the probability of the second (but be careful in determining P(E) of those individual events).

Tree diagrams are used to model potential outcomes of events based on previous events. As an example, one could model potential moves in a Chess game with a tree diagram.

Example 7.3.a: What is the probability that Luke will draw a queen followed by another queen in a standard 52 card deck?

A. $\frac{7}{52}$

B. $\frac{7}{2652}$

C. $\frac{1}{221}$

D. $\frac{1}{169}$

Example 7.3.b: Luke has 3 colors of socks, 4 shirts, and 2 pairs of pants. How many combinations can he make if he chooses 1 pair of socks, 1 shirt, and 1 pair of pants?

A. 9
B. 24
C. 20
D. 22

Solutions to Example Problems:

Solution to 7.3.a:

The answer is C. The probability of drawing the first queen is $\frac{4}{52}$ and the probability of drawing the second queen is $\frac{3}{51}$ because there are 3 queens left out of 51 possible cards left; this is an example of a dependent event. Thus, the probability of drawing a queen is: $\frac{4}{52} x \frac{3}{51} = \frac{12}{2652} = \frac{1}{221}$.

Solution to 7.3.b:

The answer is B. $3 socks\ x\ 4 shirts\ x\ 2\ pants = 24 combinations$. If you need help in answering this question, break the problem into a simpler problem: if Luke only had to match 4 shirts with 2 pairs of pants, he would have 8 possible combinations. This means that for any problem such as this, we need to multiply out each of the total number of options he has.

7.4. Connecting Theoretical And Empirical Probability

The first example below shows a connection between theoretical and empirical probability. The second example tests your knowledge of theoretical probability.

Empirical probability is an observed probability based on testing; theoretical probability is a mathematical calculation.

Example 7.4.a: Luke rolls a die 61 times. Of those 61 rolls, he rolls a six 10 times, and he rolls a four 8 times, he rolls a one 4 times. Based on Luke's observations, which number's empirical probability of being rolled is closest to its theoretical probability of being rolled?

A. One
B. Four
C. Six
D. Four and Six are equally close in theoretical and empirical probability.

Example 7.4.b: Luke tosses a die 3 times. What are the odds that he rolls a 1, then a 2, then a 4 or a 5?

A. 1/216
B. 1/128
C. 1/108
D. 1/96

Solutions to Example Problems:

Solution to 7.4.a:

The answer is C. The theoretical probability that Luke will roll any number is $\frac{1}{6}$. Of the empirical probabilities: $\frac{10}{61}$ *for* 6's; $\frac{8}{61}$ *for* 4's; $\frac{4}{61}$ *for* 1's; $\frac{10}{61}$ is the closest fraction to $\frac{1}{6}$.

Solution to 7.4.b:

The answer is C. This problem is an example of **independent** events since previous rolls of the die do not effect later rolls of the die. Let's look at the probability of each separately:

$\frac{1}{6}$: the odds of rolling a 1 on the 1st roll.

$\frac{1}{6}$: the odds of rolling a 2 on the 2nd roll.

$\frac{1}{3}$: the odds of rolling a 4 or 5 on the 3rd roll.

We multiply the three fractions because the odds of the events happening are related to the events of previous events happening; think of the situation this way: the odds of rolling a 1 on the first roll is 1/6, and OF that 1/6 times, the odds of rolling a 2 on the second roll is 1/6 and OF that one out of six rolls, the odds of rolling a 4 or a 5 after that is 1/3:

$$\frac{1}{6} x \frac{1}{6} x \frac{1}{3} = \frac{1}{108}$$

Above, I capitalized the word "of" each time to demonstrate that each "of" represents another multiplication.

7.5. Geometric probability

Geometric probability problems are generally problems such as finding the probability that of hitting a bull's eye on a dart board. Like any other probability problem, the formula for geometric probability is:

$$probability = \frac{\text{number of ways an even can occur}}{\text{total number of outcomes}}$$

In the case of the bull's eye, our formula would be:

$$probability = \frac{\text{area of bull's eye}}{\text{total area of dartboard}}$$

Example 7.5.a: A circular dartboard has a diameter of 18 inches. There are 20 numbers on a dartboard, each with an equal amount of space, and a bull's eye with a diameter of 1 inch. What is the geometric probability of hitting the number 3 on the dartboard?

A. .0492
B. .5
C. .05
D. .0498

Solutions to Example Problems:

Solution to 7.5.a:

The answer is D. First, we find the total space on the dartboard:
$\text{Area (dartboard)} = 9^2\pi = 81\pi \text{ square inches}$ where 9 is the radius of the dartboard.

Next, we need to subtract the area of the bull's eye from the total area of the dartboard because the area of the "3" on the dartboard will be $1/20^{th}$ the space of the dartboard less the area of the bull's eye. The area of the bull's eye is: $A_{bull's\ eye} = \pi(.5^2) = .25\pi\, square\, inches$ where .5 is the radius of the bull's eye.

Thus, the area of the dartboard less the area of the bull's eye is:
$81\pi - .25\pi = 80.75\pi\, square\, inches$

Next, we find the area that each of the 20 numbers occupies. Since there are 20 equally spaced areas on the dartboard, the area that "3" occupies is $Area_3 = \dfrac{80.75\pi}{20} = 4.0375\pi\, square\, inches$.

Thus, the probability of hitting the number 3 on the dartboard is:

$$P(3) = \frac{space\ of\ 3}{total\ space} = \frac{4.0375\pi}{81\pi} = .0498$$

7.6. Binomial, Uniform, And Normal Distributions

Normal distribution is represented graphically by a perfect bell curve (an example might be the height of women within a population or perhaps the grades of all students in a statistics class).

A binomial distribution will also look like a bell curve but occurs when a random sample is taken from a series of trials where there can be only 2 outcomes where each outcome has an equal probability of occurrence. An example of a binomial distribution might be an experiment where somebody tosses a coin 100 times, then records the probability, and then repeats the experiment. After the experiment is repeated a number of times, a binomial distribution will occur because there are only two possible outcomes.

In a uniform distribution, each possible outcome is just as likely to occur as another.

Note that in each type of distribution, the distribution will more perfectly represent a perfect **bell curve** as the number of samples approaches **infinity**.

Example 7.6.a: Doni arrives at his bus stop every day between 7:15 AM and 7:30 AM. Doni's bus arrives promptly at 7:30 AM each day. If the amount of time in minutes (rounded to the minute) that Doni waits for his bus is recorded each day for a year, and then plotted in a graph, which type of distribution will most accurately describe that graph?

A. A normal distribution.
B. A binomial distribution.
C. A uniform distribution.
D. Not enough information.

Solutions to Example Problems:

Solution to 7.6.a:

The answer is C. Since each possible outcome (1 minute, 2 minutes, 3 minutes, … , 15 minutes) is equally likely, the distribution will be uniform. Graphically, this might look like this:

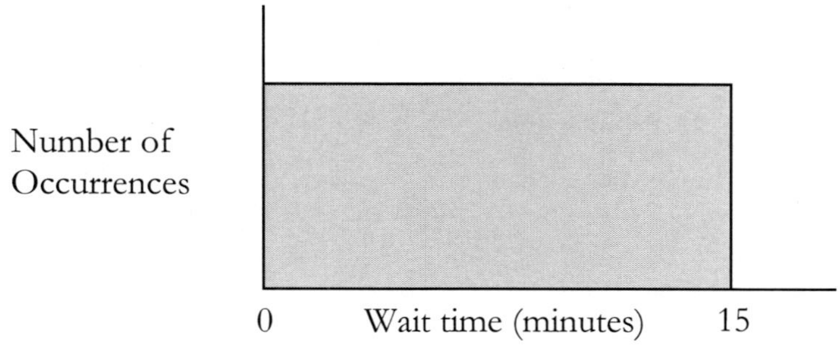

7.7. Real-World Data Collection Techniques

The use of appropriate data-collection techniques is a major theme in the study of probability and statistics, particularly in the study of random sampling and how well a certain random sample represents the population sampled as a whole. If a statistician does not use proper data-collection techniques (for example if a population is not sampled randomly), then the statistician cannot mathematically support his or her findings.

Note that as a random sample increases in size, it more accurately represents the population as a whole and therefore more accurately represents a normal distribution as a whole.

Example 7.7.a: If Luke wants to know how the people in his town feel about a certain political issue, what is the best way for him to get an accurate sample?

A. Ask 100 of his neighbors
B. Send a questionnaire to 50 people from around his town chosen randomly.
C. Send a questionnaire to 100 random teachers from around the school district?
D. Stop by the sporting goods store and ask each person who walks in (150 people).

Solutions to Example Problems:

Solution to 7.7.a:

The answer is B. A good sample has two very important characteristics: randomness and a good representation of the desired population. If Luke asks 100 of his neighbors, he has limited the data to representing whatever socio-economic level Luke's NEIGHBORHOOD has.

If Luke sends a questionnaire to 100 random teachers, the sample will effectively represent what TEACHERS in the district think about the political issue and teachers are not necessarily a representation of Luke's goal: to find out what the TOWN thinks.

Likewise, those people who go to the sporting goods store do not necessarily represent the town's population (maybe all the people who go to the sporting goods store on that particular day are hunters and therefore may have certain values in common that pertain to the political issue).

7.8. Interpreting Statistical Data

When you answer questions about charts, tables, and graphs, pay careful attention to the axes, the legend, and any graph/chart/table labels you can find.

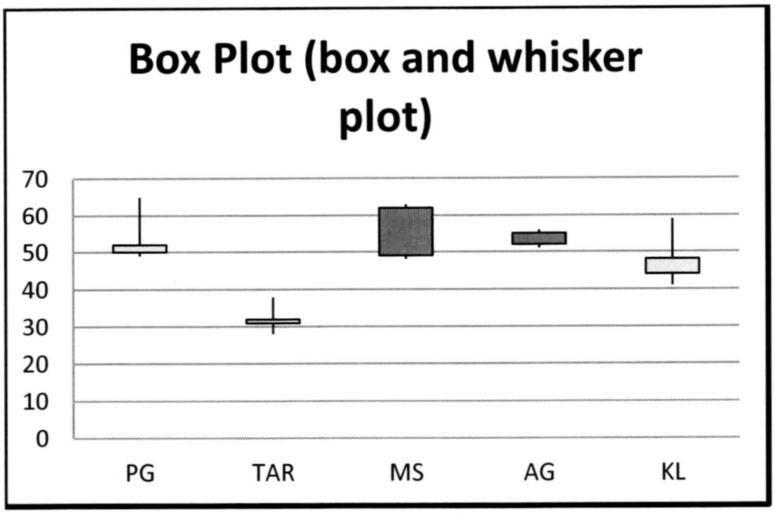

Ticker Symbol	Opening Price	High Price	Low Price	Closing Price
PG	50	65	49	52
TAR	31	38	28	32
MS	62	75	28	49
AG	55	56	51	52
KL	44	59	41	48

Example 7.8.a: The table and box plot above represent the opening price, high price, low price, and closing price of 5 stocks over the course of a day. Which stock is not accurately represented in the table above and to the right?

A. PG
B. TAR
C. MS
D. AG

Solutions to Example Problems:

Solution to 7.8.a:

The answer is C. In each of the other stocks (PG, TAR, and AG), the whiskers accurately represent the low and high prices from the table. With MS, the whiskers are way off from what they should be.

7.9. Drawing Conclusions From Data

Test writers can ask any number of questions based on sets of data. I am taking this opportunity to discuss one of the most important concepts in probability and statistics: variance.

Example 7.9.a: Which set of numbers has the highest variance?

A. {1, 2, 3, 4, 5}
B. {1, 3, 3, 3, 5}
C. {2, 5, 5, 5, 7}
D. {1, 5, 5, 5, 8}

Solutions to Example Problems:

Solution to 7.9.a:

The answer is D. Hopefully you can see from the data set that answer D has the highest variance because the data set is slightly more spread out than the other sets (I use the phrase "spread out" as a layman's term for "varied"). However, you can calculate the variance in each set by using the formula:

$$s^2 = \frac{\sum (x - \overline{x})^2}{N}$$

where s^2 is the variance, N is the population size, $\overline{x}$ is the mean of values, and x is each value in the population. I will calculate the variance of answer C below and show you the variances to the other 3 answers if you would like to practice and then check your variance-values against mine. The concept of variance is one of the most important concepts in statistics and you need to know how to calculate it for the exam.

Note that regardless of whether or not you know how to calculate variance on a calculator, the test writers will structure their questions such that you will need to know how to calculate variance, standard deviation, and mean without a calculator.

Answer C. The set is: {2, 5, 5, 5, 7}. First calculate the mean:

$$\overline{x} = \frac{2+5+5+5+7}{5} = 4.8$$

Next, calculate $x - \overline{x}$ for each term in the set:

$$x_1 = (2 - 4.8) = -2.8$$

$$x_2 = (5 - 4.8) = .2$$

$$x_3 = (5 - 4.8) = .2$$

$$x_4 = (5 - 4.8) = .2$$

$$x_5 = (7 - 4.8) = 2.2$$

Next, calculate $\sum (x - \overline{x})^2$ which is the summation of all the terms above, squared:

$$\sum (x - \overline{x})^2 = (-2.8)^2 + (.2)^2 + (.2)^2 + (.2)^2 + (2.2)^2$$

$$= 7.84 + .04 + .04 + .04 + 4.84 = 12.8$$

Next, calculate the variance by dividing 12.8 by the total number of terms.

$$the\,variance = s^2 = \frac{12.8}{5} = 2.56$$

The variance of the set in answer A is 2; the variance of the set in answer B is 1.59; the variance of the set in answer D is 4.96. You are welcome to use the answers provided to practice finding variance.

Note, if we were calculating the variance of a sample (a **subset**), rather than an entire set (a population), we would use N-1 rather than N. Likewise, if you calculate the variance of an entire **population** you use N but if you calculate variance using a subset or a sample of that population, you use N-1.

7.10. Variance, Standard Deviation, Range, Mode, Median, Mean, And Percentiles

The standard deviation of a population is found by the formula:

$$\sigma = \sqrt{\frac{\sum (x - \bar{x})^2}{N}}$$

and the standard deviation of a sample of a population is found by the formula:

$$s = \sqrt{\frac{\sum (x - \bar{x})^2}{N - 1}}$$

You may have noticed that these formulas above are simply the square roots taken of the variance of a population and of a sample of a population, respectively. You use these formulas exactly as you would use the variance formulas.

In fact, the concept of standard variation and of variance is basically the same, but the standard deviation of a data set is more useful in analyzing normalized data. Specifically, if you look at a normalized data set like IQ scores, you will see that the mean of all IQ scores is 100 and the standard deviation is always 15. This means that 68.27% of people fall between an IQ score of 85 (which is one standard deviation below 100) and 115 (which is one standard deviation above 100).

Any normalized data is set up as such: that 68.27% of the data will fall within 1 standard deviation (1 σ) of the mean (μ) and 95.45% of data will fall within 2 standard deviations (2σ) of the mean. See the figure below.

In addition, any data which falls more than 2 standard deviations from the mean is considered an **outlier** and can often times be removed from statistical analysis (outliers are often called artifacts and are considered anomalies).

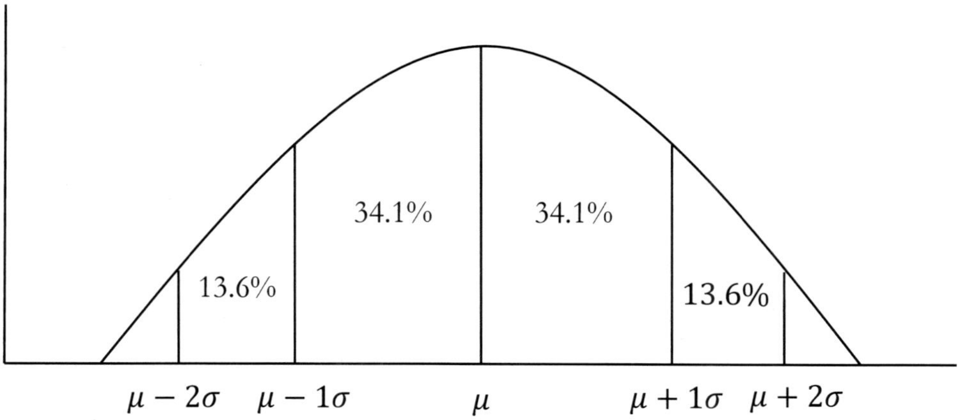

Example 7.10.a: Which is true of set A? A = {1, 2, 2, 3, 4, 4, 4, 5}

A. median > mode > mean
B. mode > mean > median
C. mode > median > mean
D. median = mode = mean

Solutions to Example Problems:

Solution to 7.10.a:

The answer is C. The mean is $\frac{1+2+2+3+4+4+4+5}{8}=3.125$; the mode, or the most common value in this group, is 4; the median, or the value which separates the higher half of the sample from the lower half of the **sample**, is 3.5; because there is an even number of numbers in the sample, we have to average 3 and 4 to find the median.

7.11. Arithmetic means and weighted averages

Average, mean, and arithmetic mean are all words for the same calculation; an average is found by adding all the values in a set and dividing by the number of items in a set.

Weighted average is a little different than arithmetic mean. A weighted average comes into play when certain values are weighed more heavily than other values.

Consider the following conceptual example: say that Luke wants to buy a new pogo-stick. Luke compares different pogo-sticks for the following attributes: jumping height, comfort, sturdiness, and weight. However, the most important thing for Luke, by far, is how high he can jump on each pogo-stick.

For Luke to simply get a value for each of these qualities and take an average would not be the best way to compare pogo-sticks; Luke must do a weighted average where he gives double or maybe triple value to how high he can jump on each pogo-stick. In weighing a certain characteristic more highly than others, Luke uses **weighted average**.

Example 7.11.a: Luke assigns the following values to a pogo stick:

Quality	Value
Sturdiness	7
Jump Height	9
Weight	2
Comfort	4

Luke weighs Jump Height three times more important than any other quality. What is the weighted average of the pogo-stick?

A. 5
B. 5.5
C. 6.67
D. 7.87

Solutions to Example Problems:

Solution to 7.11.a:

The answer is C. We can solve this by simply adding in the Jump Height three times and remembering to divide as if there were an extra two numbers:

$$\frac{Sum\, of\, Values}{Total\ Number\ of\ Values} = \frac{7+9+9+9+2+4}{6} = 6.67$$

7.12. Transformations Of Central Tendency

Mean, median, and mode are the three measurements of central tendency in statistics. The two main forms of data transformation which affect these measures of central tendency are **skewedness** and **kurtosis**. Negative skewedness creates a tail to the left of a negative distribution and positive skewedness creates a tail to the right of a normal distribution.

Positive kurtosis creates a sharper peak in a normal distribution while negative kurtosis creates a squat normal distribution curve.

Example 7.12.a: We have a sample of 4 numbers with a variance of 7. If we multiply every number by 2, what happens to the variance?

A. the variance increases
B. the variance decreases
C. nothing happens to the variance
D. impossible to tell by the given information

Example 7.12.b: Given the skewed normal distribution below, which is the mean, median, and mode?

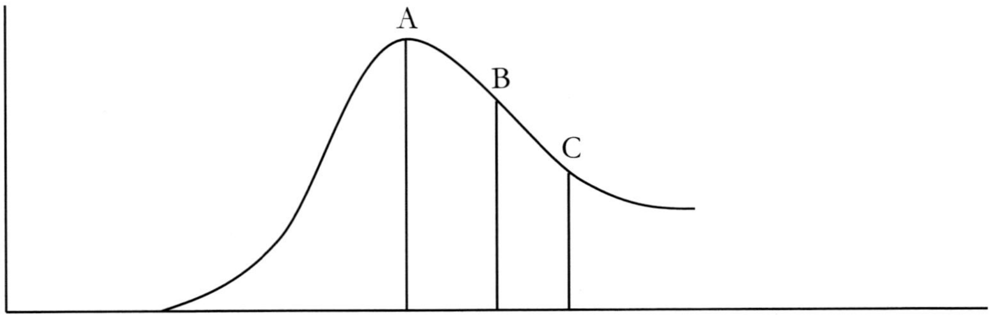

A. Mode is A, Median is B, Mean is C
B. Mean is A, Mode is B, Median is C
C. Median is A, Means is B, Mode is C
D. Mode is A, Mean is B, Median is C

Solutions to Example Problems:

Solution to 7.12.a:

The answer is A. Because the variance is 7, we know that the numbers in the sample must be different from one another (if the variance was 0, all the numbers would be the same). As an experiment, I chose to make up four numbers and look at the variance and then double all those numbers and look at the variance once again. The variance increased when I doubled the numbers in my made-up sample.

The quickest way to answer this question seems to be to use a graphing calculator to quickly test the two scenarios for variance (original numbers and doubled numbers).

Section 7.14 gives a quick guide on using a graphing calculator to find variance.

Solution to 7.12.b:

The answer is A. If you can imagine a set of data where several data points above the normalized mean are added, it may help you solve a problem such as this. When in doubt, make up some numbers to match the data transformation and see what happens to the mean, median, and mode.

7.13. Interpreting Statistical Summaries

Statistical summaries are used so that people can quickly look at a summary and then make conclusions based on that summary; in contrast, a full data set may take quite a bit of time to analyze.

Example 7.13.a: If the mean of a sample = 10, the mode = 11, the standard deviation = 4, and the range = 14, what can you say about the distribution?

A. slightly skewed left
B. slightly skewed right
C. unskewed
D. impossible to tell from the given information

Solutions to Example Problems:

Solution to 7.13.a:

The answer is A. The key information here is that the mode is slightly higher than the mean. In a graph of normalized data, the mode will be the peak of the curve regardless of whether the data is skewed or not; think about it this way: the y-axis represents number of people and the mode represents the greatest number of occurrences – therefore the mode will always be at the peak of a normal distribution.

Since the mean is slightly to the left, the left side of the peak will be slightly higher than the right side of the peak. Thus, the distribution will be slightly skewed left.

7.14. Using A Graphing Calculator In Statistics

Here is a simple guide for using a Texas Instruments calculator to perform statistical analyses:

To make a list of numbers, use the command STAT → EDIT. This lets you change the list of the numbers in L1, L2, etc.

To do an analysis on list L1, use the command STAT → CALC. The first option down is called "1-Var Stats". When you enter this command, the calculator will read "1-Var Stats" at which point you need to tell the calculator which list to analyze. Press 2nd+1 and your calculator will then read "1-Var Stats L1" since 2nd+1 is the L1 list.

Press enter when your calculator reads "1-Var Stats L1" and you will get the following readout:

$\bar{x}$ = which is the mean of your list.

$\sum x$ = which is the sum of all the numbers in your list.

$\sum x^2$ = which is the sum of all the squared numbers in your list.

Sx = which is the standard deviation if your list is a sample of a population.

σx = which is the standard deviation if your list is a full set (a population).

n = which tells you how many items you have in your list.

Example 7.14.a: What is the standard deviation of the sample set: {1, 2, 6, 8, 9, 10, 11, 14}?

A. 4.15
B. 4.25
C. 4.43
D. 4.69

Solutions to Example Problems:

Solution to 7.14.a:

The answer is C. Note that the above set is a sample rather than a population.

7.15. Correlation

Correlational relevance between two sets of data is often plotted on a single scatter plot graph with one variable plotted on the x-axis and another variable plotted on the y-axis. As an example, you can plot height on the y-axis and weight on the x-axis and then determine whether there is a correlation between height and weight based on how the scatter plot looks.

When analyzing correlational data, the **correlation** coefficient, r, is generally used. r is the slope of the line that best fits the data. The values of r are shown below:

If the scatter plotted points are so spread out that the line of best fit is horizontal, then there is no correlation between variables.

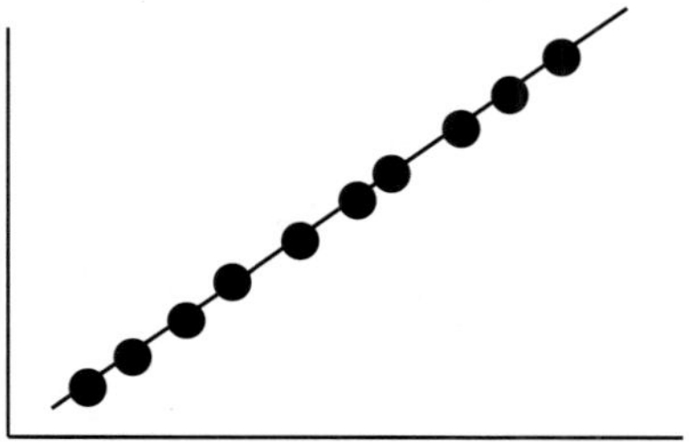

r = +1
Positive correlation

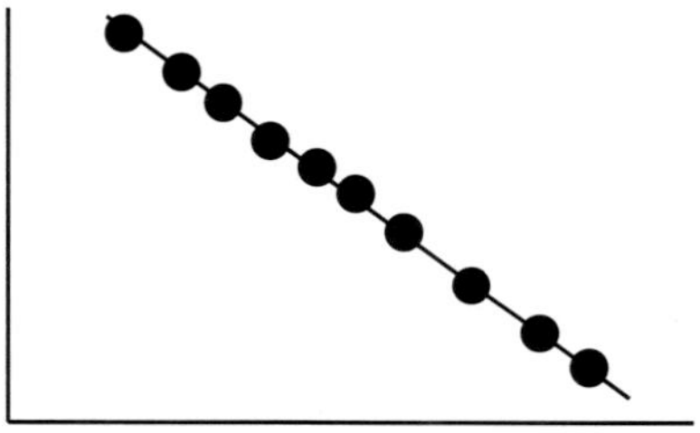

r = -1
Negative correlation

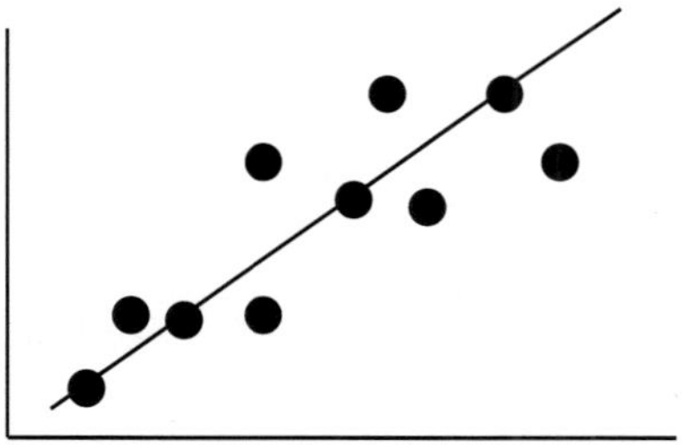

0 < r < 1
Positive correlation

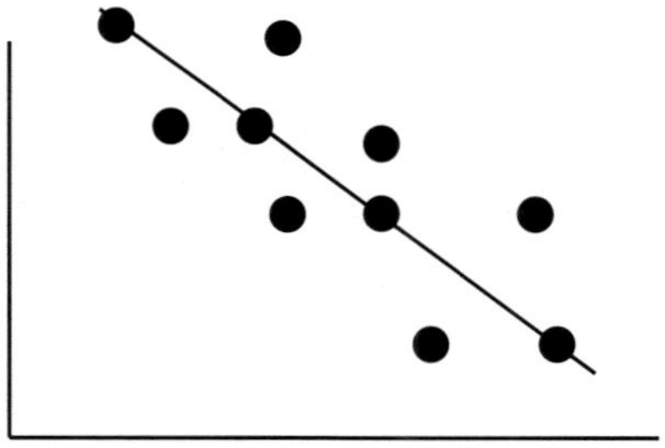

-1 < r < 0
Negative correlation

Example 7.15.a: The graph below shows Hannah's happiness vs. the amount of sunshine during the day. How would you describe the correlation between Hannah's happiness and the amount of sunshine in a day?

A. weak positive
B. strong positive
C. weak negative
D. no correlation

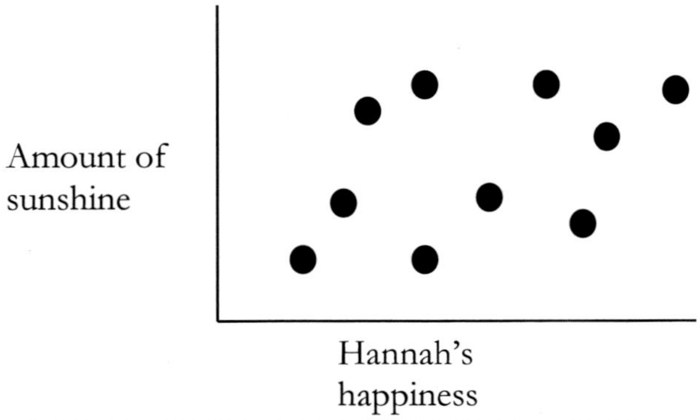

Solutions to Example Problems:

Solution to 7.15.a:

The answer is A.

The correlation is positive because the best fit line would have a positive slope ($r \approx .5$) but the correlation is a weak correlation because the plotted points do not closely hug the line of best fit.

7.16. Sampling Techniques And Sample Size

In statistics, the goal of most surveys or experiments is to find statistically significant effects; for example, if an experimenter is looking at height differences between blonds and brunettes within the Denver city limits, the experimenter needs to perform the experiment in such a way that if a height difference exists, he or she can prove that it exists statistically.

To find significant data, our experimenter needs a large enough sample size that if he or she finds a real difference in height between blonds and brunettes (maybe 1 centimeter?); the experimenter generally aims for a 5% or less chance that the data is wrong. In addition, the experimenter needs to use sampling techniques such that the population of blonds and brunettes are accurately represented in the sample (no fake blonds in this study please!)

Example 7.16.a: Which of the following have an effect on the sample size necessary for an experimenter to find an effect statistically relevant?

A. The standard deviation of the mean.
B. The confidence interval chosen.
C. All of the above.
D. None of the above.

Solutions to Example Problems:

Solution to 7.16.a:

The answer is C. Consider answer A; if the difference between both groups in an experiment is very small (the standard deviation is quite small), then you will need many more subjects in your experiment to prove a significant difference. Consider answer B; the more confident that an experimenter wishes to be that his or her results are not a **false-positive**, the more subjects are needed.

7.17. Confidence Intervals

You can use a confidence interval to indicate the reliability of an estimate within a range of values. For example, say that you have a machine that cuts lengths of wood as close to 1 foot in length as possible. The machine is not perfect; it will not cut every piece of wood exactly 1 foot long. By measuring a random sampling of cut pieces of wood, you can determine with a certain degree of confidence that each piece of wood cut will be within an estimated range (like $1\,foot \pm .25\,inches$).

Example 7.17.a: A confidence interval measures:

A. The uncertainty associated with a sampling method.
B. The uncertainty of your estimated value.
C. The probability that a certain event will occur.
D. The probability that a value will be within a certain range.

Solutions to Example Problems:

Solution to 7.17.a:

The answer is D. Answer D is a layman's definition of confidence interval.

7.18. Hypotheses

In general, a hypothesis is a proposed explanation for a phenomenon. If I wish to conduct a study on the observed phenomenon that high-fives make people happy, I may hypothesize that if high-fives make people happy, then there will be more people smiling after a high-five than frowning after a high-five.

A scientific hypothesis can be tested using the scientific method. Also, a hypothesis must be falsifiable (can be disproven). A **null hypothesis** suggests that there is no relationship between observed phenomena; as per our example, here is my null hypothesis: there will not be significantly more people smiling after a high-five than frowning after a high-five.

A **type I error** occurs when a statistical test rejects a true null hypothesis. Think **false-positive**. For example, if there really is no phenomenon that high-fives make people happy and, by chance, during our testing, I observe the rare people that are made happy by high-fives, then a type I error has occurred. Note that in this example, a type I error seems counter intuitive because high-fives probably do make people happy. A type I error is usually represented in statistics by the Greek letter α.

A type II error occurs when there is a failure to reject a false null hypothesis. Think **false-negative**. In our case, a **type II error** would occur if we failed to reject the null hypothesis that high-fives do not make people happy (when they really do make people happy). In other words, we find that high-fives don't make people happy, but they really do. A type II error is represented in statistics by the Greek letter β.

Example 7.18.a: When a pregnancy test determines that a woman is pregnant but she is really not, what type of error has occurred?

A. A type I error, α
B. A type II error, β
C. A type I error, β
D. A type II error, α

Solutions to Example Problems:

Solution to 7.18.a:

The answer is A. This is a false positive – a type I error.

Chapter 8

Calculus and Discrete Mathematics

8.1. Series, Sequences, and Limits of Functions

Picture yourself out to dinner at the end of a delicious meal. You and your significant other ordered the most delicious chocolate cake and there is only one bite left. You pick up your fork and cut that bite in half. Your date takes out his or her fork and bites the remaining half in half. This continues, with each person taking half of the remaining delicious chocolate cake over and over and over again forever. In mathematics, we would say that the size of the cake approaches zero as the **limit** (of the number of bites) approaches infinity.

You will need to understand how limits work. $\lim_{x \to a} f(x)$ (read as "the limit of *f* of *x* as *x* approaches *a*") is the general format for reading a limit. I have included several limit problem examples throughout this chapter; it may be easiest to relearn limits through trial and error.

As for series, which are primarily taught in Calculus 3, you must first learn the nomenclature. As an example, $\sum_{k=0}^{5} x$ is the sum of the numbers 0 through 5. You work a sum by plugging in the number on the bottom of the upper case sigma into the function (in this case, x is the function) and then adding the next term which is the next integer after 0 plugged into the function. You continue in this manner until you have added the last integer in the series (in this case 5 is the highest integer). The calculation looks like this: 0 + 1 + 2 + 3 + 4 + 5.

An **infinite sequence** is a list of numbers that goes on forever. You may see an infinite sequence of numbers like $\frac{1}{2}, \frac{2}{3}, \frac{3}{4}, \ldots, n$ or an infinite sequence like 1, 2, 3, 4, … , *n*. If a sequence moves toward a finite number as the terms of the series approaches infinity, then the sequence is said to **converge**; for example, $\frac{1}{2}, \frac{2}{3}, \frac{3}{4}, \ldots, n$ converges toward 1 (think of the 500th term). If a sequence does not approach a finite number, it is said to **diverge**. The sequence $1, 2, 3, \ldots, n$ diverges because it approaches infinity as the number of terms approaches infinity.

Example 8.1.a: Evaluate: $\lim_{x\to\infty}\left(\frac{1-\cos(x)}{x^2}\right)$

A. Does not exist.
B. $-\infty$
C. ∞
D. 0

Solutions to Example Problems:

Solution to 8.1.a:

The answer is D. The limit of the top oscillates between 0 and 1 while the bottom becomes infinity. Thus the limit approaches 0.

8.2. Asymptotes

When the graph of a function approaches a line without ever touching it even though the function will get infinitely close to that line, that line is an **asymptote**. For the most part, the asymptotes that you will find on the exam will be vertical or horizontal. You can find asymptotes simply by graphing and looking for them. For example, graph the function $f(x)=1/x$ and you should see a vertical asymptote at $x = 0$ and a horizontal asymptote at $y = 0$.

Example 8.2.a: What are the asymptotes of the graph created by the function

$$f(x)=\frac{1}{x^2-4}$$

A. f(x) = 0; x = -2; x = 2
B. f(x) = -2; f(x) = 2; x = 0
C. x = 4; x = -4
D. f(x) = 0; x = 4; x = -4

Solutions to Example Problems:

Solution to 8.2.a:

The answer is A. Graphing the function on a graphing calculator, you can see that there is one horizontal asymptote at f(x) = 0 and two horizontal asymptotes: one at x = 2 and another at x = -2.

8.3. Continuity and Discontinuity

Discontinuity occurs when certain values for x result in one of two things: a zero in a denominator or a negative in a square root. For example, discontinuity will occur when $x = 0$ in the function $f(x) = \frac{1}{x}$.

Use the graph of $f(x)$ below to answer Example 8.3.a:, Example 8.3.a:, and Example 8.3.b:.

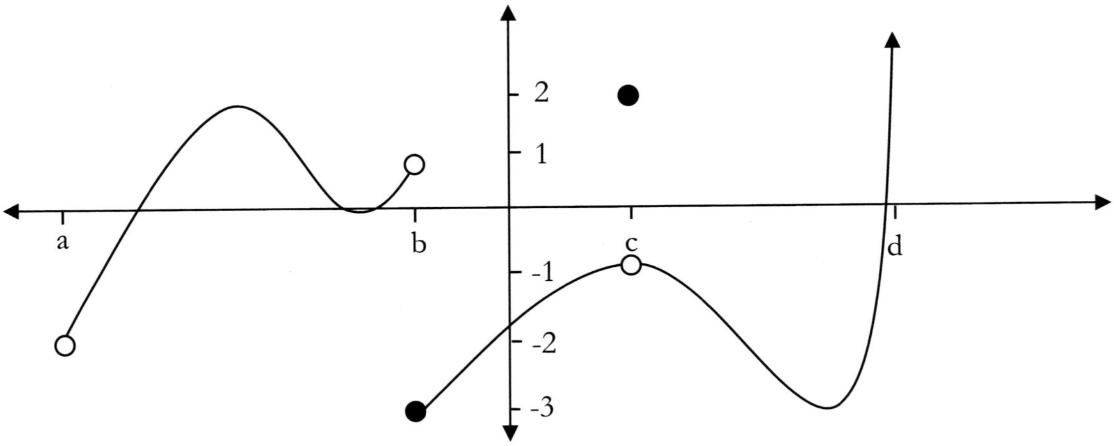

Example 8.3.a: Which of the following answers contain all parts which DO NOT exist?

A. $\lim\limits_{x \to a}$; $f(b)$; $\lim\limits_{x \to c^+}$

B. $\lim\limits_{x \to b^-}$; $\lim\limits_{x \to d}$; $\lim\limits_{x \to a}$

C. $\lim\limits_{x \to b}$; $\lim\limits_{x \to d}$; $\lim\limits_{x \to a}$

D. $\lim\limits_{x \to d^-}$; $\lim\limits_{x \to a}$; $f(c)$

Example 8.3.b: $f(c)-\lim_{x\to c}+\lim_{x\to b^{+}}=$

A. 0
B. 1
C. 3
D. 5

Example 8.3.c: What kind of discontinuity exists at point c?

A. removable
B. jump
C. infinite
D. none, f(x) is continuous through point c

Solutions to Example Problems:

Solution to 8.3.a:

The answer is C. Since the limit as x approaches b from the left and from the right are different, that limit does not exist; for the same reason, the limit as x approaches d and as x approaches a do not exist either.

Solution to 8.3.b:

The answer is A. This is because f(c) is 2, $\lim_{x \to c}$ is -1, and $\lim_{x \to b^+}$ is -3. Thus, 2-(-1)-3=0.

Solution to 8.3.c:

The answer is A. f(b) is an example of jump discontinuity. Infinite discontinuity does not exist in the graph above.

8.4. Average Rates of Change

The average rate of change is the slope between two distinct points. The average rate of change between two points is also called a **secant line.** In contrast, finding the slope at a single point requires the use of a derivative or a limit; slope at a single point is called a **tangent line.**

Average slope between two points is simply rise over run or change in y over change in x. Don't get confused by symbols which represent "change." These symbols can be: dx, dy where dx represents the change in x and dy represents the change in y. "d" is the Greek letter delta and stands for "change."

Example 8.4.a: A car's **displacement** as a function of time is represented by the equation $s = t^2 - 4t + 2$ where t is measured in seconds and s is measured in meters. Find the average velocity between t = 1 and t = 4.

A. -1 m/s
B. 1 m/s
C. 2 m/s
D. 4 m/s

Solutions to Example Problems:

Solution to 8.4.a:

The answer is B. This problem is asking for a slope between the points where t = 1 and where t=4. When you see a practical problem like this, sketch it out and label the axes; you will see that the x axis is seconds and the y axis is meters (s stands for displacement). Anyway, at t = 1, the displacement in meters is $s(1)=1^2 -4(1) +2 = -1$. Thus, the first point we are looking for is (1,-1).

At t = 4, the displacement in meters is $s(4)= 4^2 -4(4) + 2 = 2$. Thus, the second point we are looking for is (4, 2). The slope between the points (1-1) and (4, 2) is 3/3 or 1 meter per second.

8.5. Derivatives of Functions and Limits of Slope

One part of the fundamental theorem of calculus describes finding the slope of a line at a single point. The equation for finding the derivative as a limit is:

$$\lim_{h\to 0}\frac{f(x+h)-f(x)}{h}$$

where h is the change in x. To visualize this, think of a continuous function with two points on it. The difference between those two points in the x dimension is h; as the difference between the two points gets infinitely close to 0, the secant line will become the slope of a single point. This concept is a major theme in Calculus; you will definitely see questions about the fundamental theorem of calculus on the exam. I included a second description in the answer to the following example.

Example 8.5.a: It is possible to find the instantaneous rate of change of a function by:

A. taking the slope of function where the change in x = 0.
B. taking the slope of a function where the change in y = 0.
C. taking the slope of a function where the limit of $dx \to 0$.
D. taking the slope of a function where the limit of $dy \to 0$.

Solutions to Example Problems:

Solution to 8.5.a:

The answer is C. The slope of a function is $\frac{rise}{run}$ or $\frac{dy}{dx}$. As the change in x (in this case the change in x represents a measurement in time) becomes 0, the slope becomes instantaneous. To understand this visually, draw a curve and two points on the curve. Draw a line between those points; this line is a secant line. Now make the points closer together (dx gets smaller). As the distance between the two points gets infinitely close to zero, the line represents a tangent line which is the slope of a line at a single point and also known as a derivative; this concept is the main idea in differential calculus.

Answer A is a trap answer; you cannot evaluate a slope where x is exactly equal to 0.

8.6. Derivatives are Instantaneous Rates of Change

A derivative is an instantaneous rate of change. The word problems below describe a car's displacement. The rate of change of the displacement of the car at a specific point in time (instantaneous rate of change) is its velocity and the rate of change of velocity at a specific point in time is the acceleration of the car.

In general, to find out what the derivative of a function might represent in the context of a word problem, take the y axis over the x axis. As an example, if a car's displacement is measured in meters on the y-axis of a graph and time in seconds is represented on the x-axis, then the $\frac{y-axis}{x-axis}$ is $\frac{meters}{second}$ or velocity. In the same manner, the derivative of graph which shows velocity on the y-axis and time on the x-axis would be the slope or velocity or $\frac{meters}{second^2}$ or acceleration.

Example 8.6.a: A car's displacement as a function of time is represented by the equation $s = t^2 - 4t + 2$ where t is measured in seconds and s is measured in meters.

Find the velocity at t = 3 seconds of a car where

A. -1 m/s
B. 1 m/s
C. 2 m/s
D. 4 m/s

Example 8.6.b: The figure below shows displacement of a particle over time where positive displacement, s, represents position to the right of the starting point. When is the particle moving left?

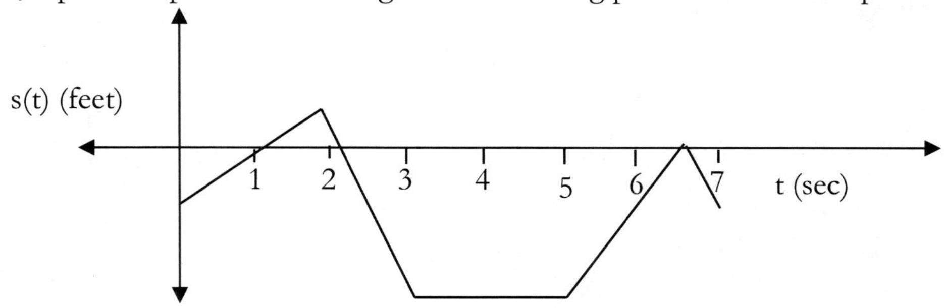

A. [0, 2] ∪ [5, 6.5]

B. [0, 2.5] ∪ [6, 7]

C. [2.5, 6]

D. [2, 3] ∪ [6.6, 7]

Solutions to Example Problems:

Solution to 8.6.a:

The answer is C. This question is asking you to find instantaneous velocity, which requires that we take a derivative of $s = t^2 - 4t + 2$. Taking the first derivative of s, we get $v = s' = 2t - 4$; and plugging in 3 for t we have $s'(3) = 2m/s$.

Solution to 8.6.b:

The answer is D. Since the y-axis of the graph above represents displacement, any time the graph has a downward slope, the particle is moving left. Thus the answer is [2, 3] and [6.6, 7].

8.7. Investigating Behavior Using Derivatives

Here is a brief synopsis on finding derivatives:

Given that $f(x) = ax^b$, taking the first derivative uses the formula:

$$f'(x) = bax^{b-1}$$

For example, start with a function, say $y = x^2 + 3x + 1$. Using the formula above, the derivative would be:

$$y' = (2)x^{2-1} + (1)x^{1-1} + (0)1x^{0-1}$$
$$\text{or } y' = 2x + 3$$

Notice that I inserted an x^0 next to the +1 to demonstrate that the formula applies to coefficients and x-terms alike.

Critical points are points on a function where the slope changes from positive to negative, or the slope changes from negative to positive. If the slope of a function is 0 at a specific point then its slope has changed from positive to negative or vice-versa. To find critical points, take the derivative of a function and set it equal to 0.

You may see questions concerning **concavity** on the exam. If a function is concave up, then the function's second derivative will be greater than 0. Likewise, a function's second derivative will be negative wherever the function is concave down.

Example 8.7.a: Determine the concavity of the function $f(x) = -2x^3 + 2x + 5$ at the point x = 4 and any find any **critical points**.

A. $-\frac{1}{\sqrt{3}}\ and + \frac{1}{\sqrt{3}}$; concave down

B. $-\frac{1}{\sqrt{3}}\ and + \frac{1}{\sqrt{3}}$; concave up

C. $-\frac{1}{3\sqrt{3}}\ and + \frac{1}{3\sqrt{3}}$; concave down

D. $-\frac{1}{3\sqrt{3}}\ and + \frac{1}{3\sqrt{3}}$; concave up

Solutions to Example Problems:

Solution to 8.7.a:

The answer is A. To find the critical points of the function, take the first derivative and set it equal to 0:

$$f'(x) = -6x^2 + 2$$

$$0 = -6x^2 + 2$$

$$-2 = -6x^2$$

$$\frac{1}{3} = x^2$$

Thus, you can see that the critical points of the function – the points where the slope of the function will be 0- are at $-\frac{1}{\sqrt{3}}$ *and* $+\frac{1}{\sqrt{3}}$.

To determine the concavity, we find the second derivative and plug in 4:

$$f''(x) = -12x$$

$$f''(4) = -48$$

Since the second derivative of the function at 4 is a negative number, the slope is constantly decreasing around the point x=4 which we know is a minimum or maximum. Visually, this will be a maximum of the function and will be concave down. If you prefer, graph the original function and you can see that the function is concave down at x = 4.

8.8. Analysis through Differentiation

When the slope of a line changes from increasing to decreasing, it does so at an **inflection point**. I made a figure below to demonstrate inflection points, minimums, maximums, and critical points as its easier shown than explained.

To find an inflection point, the point where a slope changes from increasing to decreasing, use the 2nd derivative. Once you take a 2nd derivative of a function, plug in the point. If you get a positive result, then the slope is increasing; if you get a negative result, the slope is decreasing. Inflection points can be found by setting the 2nd derivative of a function to zero and solving.

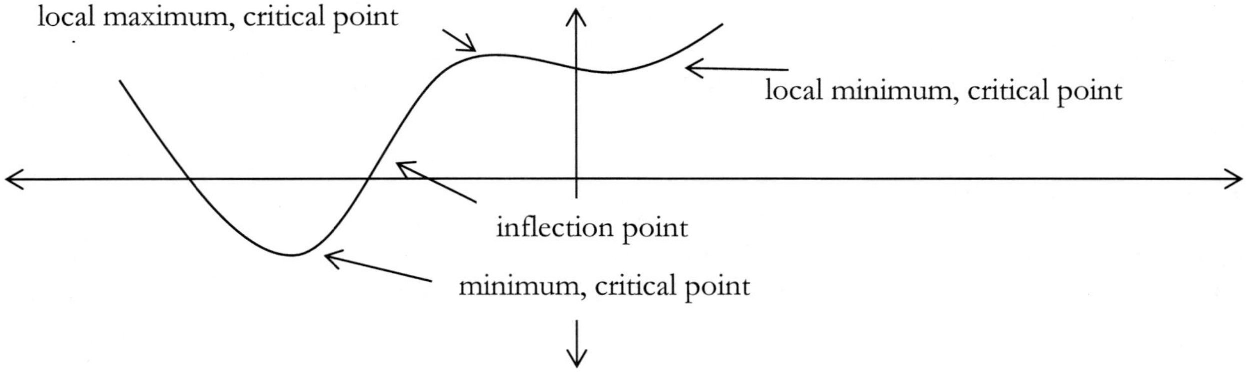

Example 8.8.a: A car's displacement as a function of time is represented by the equation $s = t^2 - 4t + 2$ where t is measured in seconds and s is measured in meters. Find the acceleration of the car at t = 2.

A. -1 m/s^2

B. 1 m/s^2

C. 2 m/s^2

D. 4 m/s^2

Solutions to Example Problems:

Solution to 8.8.a:

The answer is C. The acceleration can be found by taking the second derivative of $s = t^2 - 4t + 2$ and plugging in t = 2.

Thus, $a = v' = s'' = 2m/s^2$.

Note that the acceleration is constant in this problem (you can see that there is no place to plug t in).

8.9. Approximating the Area under a Curve

You can often solve area-under-curve problems by splitting the area into basic shapes like triangles and semi-circles. Notice that the graph below is not perfectly accurate; inaccurate drawings are fair game on the exam.

Example 8.9.a: What is the area between the two dashed lines given that the two dashed lines are the functions $f(x) = x^2$ *and* $g(x) = 4$?

A. 32/3
B. 11
C. 34/3
D. not enough information given

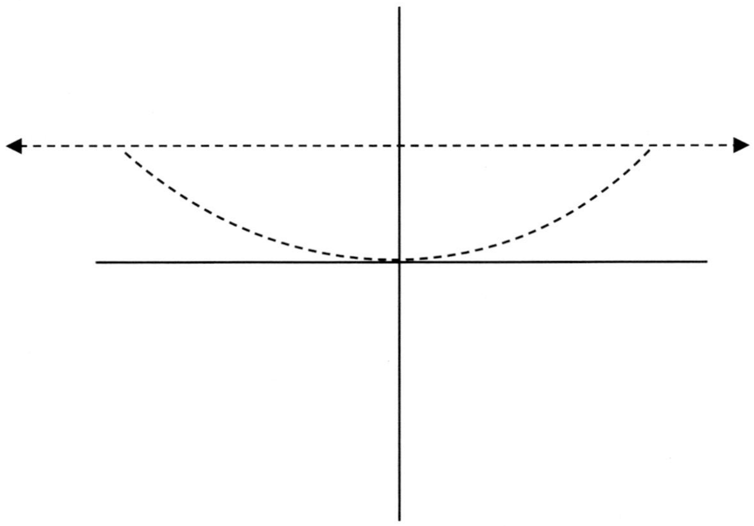

Solutions to Example Problems:

Solution to 8.9.a:

The answer is A. First, to find the points where g(x) and f(x) meet, plug in 4 for the y-value in f(x) and solve for x; you should find that when x = +2, and x = -2, $f(x) = x^2 = 4$.

Next, I calculated using integrals. There are a couple methods you can use to do this problem. The easiest might be to realize that the area is a rectangle between the lines y=0, y=4, x = -2, and x=2 minus the area under the curve $f(x)=x^2$ between the lines x = -2 and x = 2.

Therefore, we have the area of the rectangle 4 x 4 = 16 minus $\int_{-2}^{2} x^2$. Thus, Area = 16 - $\frac{x^3}{3} |_{-2}^{2}$ (read x cubed over three evaluated between -2 and 2). Area = 16 - $\frac{2^3}{3} - \frac{-2^3}{3} \Rightarrow Area = 16 - \frac{8}{3} - \frac{-8}{3} \Rightarrow$ $Area = 16 - \frac{16}{3} \Rightarrow Area = \frac{48}{3} - \frac{16}{3}$. Thus, Area = 32/3.

8.10. Definite Integrals

Given $\int_{a}^{b} f(x)dx$, the definite integral of a function is the area between f(x) and the x-axis between the x-coordinates a and b. Any area below the x-axis will be subtracted from the overall area. The difference between an indefinite integral and a definite integral is that with an indefinite integral, a and b are not defined and, thus, one cannot find the actual area between a function and the x-axis. One can, however, use an indefinite integral for many purposes – most of which are for theoretical problems.

Example 8.10.a: Which integral represents the shaded area below?

A. $\int_0^3 (x-1)\,dx$

B. $-\int_0^1 x dx + \int_1^3 x dx$

C. $1 + \int_0^3 x\,dx$

D. Two of the above

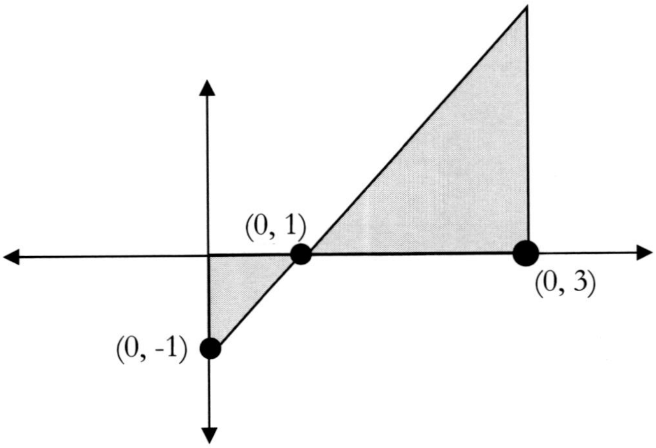

Solutions to Example Problems:

Solution to 8.10.a:

The answer is A. The function $f(x) = x - 1$ represents the graph accurately and the domain of evaluation from 0 to 3 is also accurate to the graph drawn.

8.11. Connecting Integration and Differentiation

You can move back and forth between a function and its antiderivative by integrating and differentiating if that function is differentiable on a given interval.

Example 8.11.a: Find $\int_a^b \frac{d}{dx}(f(x))\,dx$ given that $f(x)$ is continuous and defined on [a, b].

A. $[f'(x)]_a^b$

B. $f(b) - f(a)$

C. $f'(b) - f'(a)$

D. 0

Solutions to Example Problems:

Solution to 8.11.a:

The answer is B. The **antiderivative** of $\frac{d}{dx}(f(x))dx$ is simply $f(x)$. If we evaluate $\frac{d}{dx}(f(x))dx$ between a and b we get $[f(x)]_a^b$ or $f(b)-f(a)$. Note that the most difficult part of this problem is realizing that the integral and the differential d/dx counteract each other. Also note that if we do not know whether the function is differentiable on the interval [a, b], then we cannot solve this problem.

8.12. Real-World Integration

Integration of a function determines the area between the function and the x-axis. To relate this to real life problems, consider a simple example: say that a car is traveling at a constant velocity of 50 miles/hour for 5 hours. The car travels 250 miles total. If you graph the example, you can see that 250 is the area created by multiplying the vertical and horizontal dimensions. To determine the units of your integration, multiply your two axes and keep track of units. In this example, $\frac{miles}{hour} x\, hours = miles$.

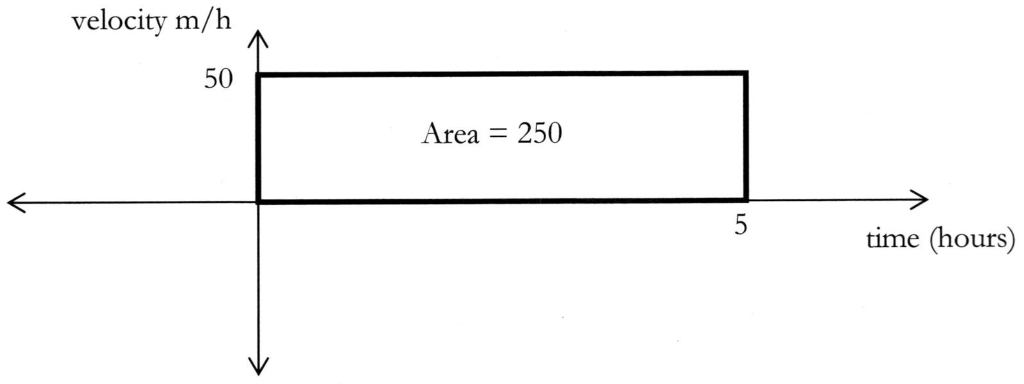

Example 8.12.a: Between the hours of 7pm and 11pm, students enter prom at a rate of $p = t^2 + 10$ where 7pm = t_0 = 0. How many people enter prom between 8pm and 11pm?

A. 51
B. 47
C. 53
D. 41

Solutions to Example Problems:

Solution to 8.12.a:

The answer is A. This can be solved using the integral $\int_1^4 t^2 + 10dx$. Notice that in the equation, 1 represents 8pm and 4 represents 11pm because they are 1 and 4 hours after 7pm, respectively. Solving, we get $\frac{t^3}{3} + 10t \mid_1^4$.

Thus:

$$\left[\frac{(4)^3}{3} + 10(4)\right] - \left[\frac{(1)^3}{3} + 10(1)\right] = \left[\frac{64}{3} + \frac{120}{3}\right] - \left[\frac{1}{3} + \frac{30}{3}\right] = \left[\frac{184}{3}\right] - \left[\frac{31}{3}\right] = \frac{153}{3} = 51\,people.$$

8.13. Recursive Formulas

Recursion is the process of taking an initial term and then repeatedly applying a process to each successive term. For recursion, one always needs to know the previous term in order to find the next one.

A **difference equation** is essentially the equation which defines the process of how each successive term is found in a recursive formula.

For example, say you want to find the 3rd term in a sequence defined by the recursive formula:
$t_1 = 3\,and\,t_n = (t_{n-1})^2$

The second term would be: $t_2 = (t_{2-1})^2 = (t_1)^2 = (3)^2 = 9$

The third term would be: $t3 = (t_{3-1})^2 = (t_2)^2 = (9)^2 = 81$

Example 8.13.a: Find the 5th term in a sequence defined by the recursive formula:
$t_1 = 2\,and\,t_n = 5t_{n-1}$

A. 250
B. 500
C. 1000
D. 1250

Solutions to Example Problems:

Solution to 8.13.a:

The answer is D.

$t_1 = 2$

$t_2 = 5(t_{2-1}) = 5t_1 = 10$

$t_3 = 5(t_{3-1}) = 5t_2 = 50$

...

$t_5 = 1250$

8.14. Graph Theory

Graphs of graph theory and graphs of functions are not the same concepts. **Graph theory** is a theme of discrete mathematics and involves modeling objects in a collection of objects. When mathematicians model processes or systems, they often use a graphical model to visually represent a system or a natural phenomenon. For example, a mathematician may model species movement where each vertex in a graphical representation represents a region and each edge (or connecting line) represents a migratory movement.

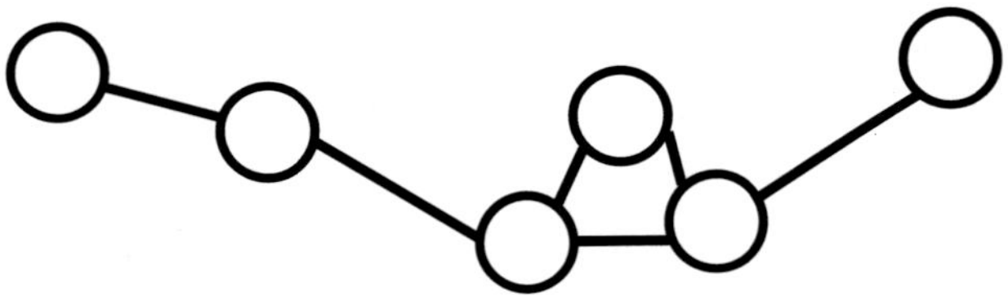

Example 8.14.a: How many edges does a three-dimensional figure with 8 vertices have? Each of the vertices has a degree of 3?

A. 12 edges
B. 8 edges
C. 24 edges
D. 14 edges

Solutions to Example Problems:

Solution to 8.14.a:

The answer is A. When the problem states that the vertices each have a degree of 3, it means that 3 edges approaches each **vertex**. If you sketch this out, you may be able to tell that the 3 dimensional figure is a cube. Thus, there are 12 edges.

8.15. Pascal's Triangle

Pascal's triangle is a triangular representation of numbers where the two numbers above a number will add to the number below them. Here is a representation:

1
1 1
1 2 1
1 3 3 1
1 4 6 4 1
1 5 10 10 5 1

And so forth. Pascal's triangle is one of those fantastic patterns that shows up in many mathematical processes, but people rarely actually use. The only way that I have ever seen Pascal's triangle used in middle school mathematics is in binomial expansion with advanced 8th graders. In the expansion below, notice how the coefficients in the binomial expansion follows Pascal's triangle:

$$(x+y)^0 = 1$$

$$(x+y)^1 = 1x + 1y$$

$$(x+y)^2 = 1x^2 + 2xy + 1y^2$$

$$(x+y)^3 = 1x^3 + 3x^2y + 3xy^2 + 1y^3$$

Example 8.15.a: What is the binomial expansion for $(x+1)^5$?

A. $1x^5 + 4x^4 + 6x^3 + 4x^2 + x$

B. $1x^5 + 4x^4 + 6x^3 + 4x^2 + 1$

C. $1x^5 + 5x^4 + 10x^3 + 10x^2 + 5x^2 + 1$

D. $1x^5 + 5x^4 + 10x^2 + 1$

Solutions to Example Problems:

Solution to 8.15.a:

The answer is C. Pascal's triangle is useful here because doing a binomial expansion to a 5th power takes quite a bit of time. Note that the very first row of Pascal's triangle will represent $(x+1)^0$. Thus, the 6th row down which is 1 5 10 10 5 1 will be the coefficients of $(x+1)^5$.

8.16. Permutations and Combinations

There are 4 main types of permutations and combinations that you will have to deal with on the exam.

Permutations with repetition: n^r where n is the number of items to choose from and you choose r of them.

Permutations without repetition: $\dfrac{n!}{(n-r)!}$

Combinations with repetition: $\dfrac{(n+r-1)!}{r!(n-1)!}$

Combinations without repetition: $\dfrac{n!}{r!(n-r)!}$

To start, if order is important, like the combination of a lock or a license plate number, then you use a **permutation**; if order is not important, use a **combination**.

If items can be repeated, like letters on a license plate, use the "with repetition" formula; if items cannot be repeated, like if you have to select committee members and each person can only have 1 job, then use the "without repetition" formula.

Example 8.16.a: Find the number of ways to choose 3 pieces of fruit out of a selection of 8 different kinds of fruit where each kind of fruit can only be chosen once.

A. 24 ways
B. 42 ways
C. 56 ways
D. 72 ways

Solutions to Example Problems:

Solution to 8.16.a:

The answer is C. If the order does not matter, we use the formula for a combination; if the order does matter, we use the formula for a permutation. In the case above, the order does not matter. Also, there can be no repetition because once you pick a piece of fruit, that piece of fruit is gone from the selection. Thus, the formula is $\frac{n!}{r!(n-r)!}$ where n is the total number of kinds of fruit and r is the number of pieces we choose. Thus, we have:

$$\frac{8!}{3!(8-3)!}=\frac{8!}{3!5!}=\frac{8x7x6x5x4x3x2x1}{3x2x1x5x4x3x2x1}=\frac{8x7x6}{3x2x1}=\frac{336}{6}=56.$$

8.17. Mathematical Induction

Mathematical induction is a method of proof used in mathematics to show that a certain statement is true for all natural numbers. To do proof by mathematical induction, you assume that a theorem is true for any natural number, k.

You must then make an inductive step by showing that if the statement is true for k, or any natural number, then the statement will also be true for the following natural number, k + 1. In doing so, you show that the next natural number also holds true for the statement and thus every successive natural number will also prove true (it's like a domino effect).

Example 8.17.a: Consider the general statement: $1 + 2 + 3 + \ldots + n = \frac{n(n+1)}{2}$ is true for any natural number. To prove the general statement by mathematical induction, first we assume that the statement is true for any natural number, k:

$1 + 2 + 3 + \ldots + k = \frac{k(k+1)}{2}$

Which of the following represents the statement that we must prove?

A. $1 + 2 + 3 + \ldots + k + (k+1) = \frac{k(k+1)}{2}$

B. $1 + 2 + 3 + \ldots + k + (k+1) = \frac{k(k+2)}{2}$

C. $1 + 2 + 3 + \ldots (k+1) = \frac{(k+1)(k+2)}{2}$

D. $1 + 2 + 3 + \ldots + k(k+1) = \frac{k(k+1)(k+1)}{2}$

Solutions to Example Problems:

Solution to 8.17.a:

The answer is C. Showing that an expression is true for k = 1 and consequently that the expression is true for k+1 is the definition of mathematical induction; to do so, we replace each k with k+1.

One major advantage that you have on the exam is that because the Praxis II ® (0061) exam is scored via Scantron machine, they will not make you actually prove any theorems on the exam.

8.18. Modeling through Sequences and Series (Not Male Modeling)

As discussed earlier in this chapter, a sequence is a list of numbers; a series is a summation of numbers. Often a real world problem can be modeled using a sequence or a series. Sequences and series are major themes of Calculus III, but also show up in basic algebra in the form of triangular numbers and geometric numbers.

A **triangular number** represents the number of objects in an equilateral triangle, like this:

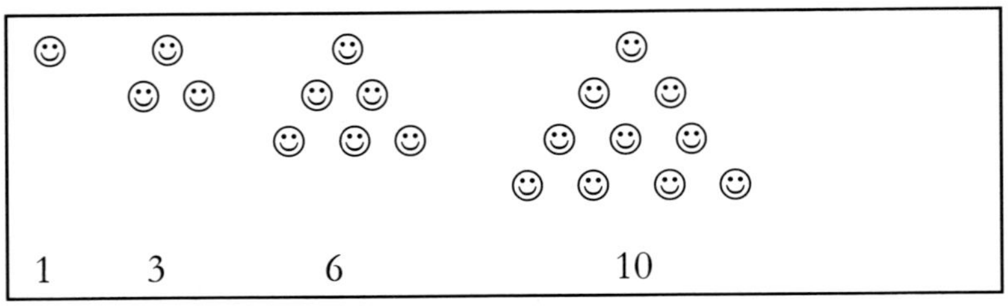

Each equilateral triangle is made up of smiley faces. Thus, the set of triangular numbers is the set $\{1, 3, 6, 10, 15, 21, \ldots\}$ or $\frac{n(n+1)}{2}$.

Any other geometric number will follow a pattern of progression. The main concept to gleam from this topic is that you may be required to write algebraic expressions from given patterns.

Example 8.18.a: A pyramid has 40 blocks on its lowest level in a 10x4 grid. If each ascending level of the pyramid has 1 less block in its width and in its length, which Riemann sum represents the total number of blocks?

A. $\sum_{k=1}^{4}(1+k)(7+k)$

B. $\sum_{k=0}^{3}(1+k)(7+k)$

C. $\sum_{k=0}^{3}(10-k)(4-k)$

D. Two of the above.

Example 8.18.b: Doni creates the figures below using toothpicks where the firs toothpick. What algebraic expression best represents the number of toothpicks figure number, f, where figure 0 does not exist?

Figure 1 Figure 2 Figure 3 Figure 4

A. $t = 3f - 2$

B. $t = f + 3$

C. $t = 2f + 3$

D. $t = 3(f - 1)$

olutions to Example Problems:

Solution to 8.18.a:

The answer is D. The lowest level of the pyramid has 10x4 = 40 blocks.

The 2nd lowest level has 9x3 = 27 blocks.

The 3rd lowest level has 8x2 = 16 blocks.

The highest level has 7x1 = 7 blocks.

Thus, the total number of blocks equals $40+27+16+7=90\,blocks$. You can approach this problem in many ways, but it may be easiest to calculate each sum and see which ones equal 90. Answer A has $(1+1)(7+1)+(1+2)(7+2)+(1+3)(7+3)+(1+4)(7+4)$ blocks which does not add to 90.

Answer B has $(1+0)(7+0)+(1+1)(7+1)+(1+2)(7+2)+(1+3)(7+3)$ which DOES add to 90.

Answer C has $(10-0)(4-0)+(10-1)(4-1)+(10-2)(4-2)+(10-3)(4-3)$ which also equals 90. Thus, two answers are correct.

Solution to 8.18.b:

The answer is A. Here is the pattern as described in a table:

Toothpicks	?	1	4	7	10
Figure Number	0	1	2	3	4

My favorite way to create algebraic expressions is to start with a table. Next, I look to see if my table represents a linear progression by whether or not my increases in slope (change in y over change in x) remain constant.

In the problem above, my slope remains constant: $\dfrac{change\,in\,y}{change\,in\,x}=\dfrac{toothpicks}{figure\,number}=\dfrac{3}{1}$. In other words, the slope is 3. Next, look to see what the y-intercept would be by seeing what your y-value would be if your x were 0. If you follow the pattern down, you can see that the y-intercept should be -2; of course, -2 toothpicks makes absolutely no sense at all, but the question tells us to ignore Figure 0. Thus, the equation for our linear equation is: $t=3f-2$.

Note that you could simply work backwards by plugging in the figure number and looking to see if you get the right amount of toothpicks for every figure.

Chapter 9

Formulas

The official Praxis II ® (0061) Exam study guide suggests that you know how to use a few specific formulas. The best way for me to explain how to use the formulas is through a series of examples.

9.1. Formula for the Volume of a Right Cone or a Pyramid

$$V = \frac{1}{3}Bh$$

where h stands for the height from the apex of the pyramid (or cone) straight to the base and B stands for the area of the base.

Example 9.1.a: Find the length of the base of a regular square pyramid with a base width of 5 and a height of 7 and a volume of 70 cubic units.

A. 4
B. 5
C. 7
D. 6

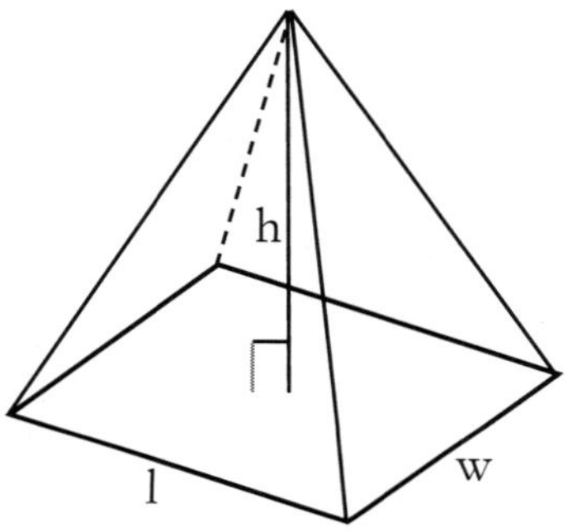

Solutions to Example Problems:

Solution to 9.1.a:
The answer is D. First, we must split the B from the formula into length times width. The formula becomes $V = \frac{1}{3} wlh$ since the area of the base is length times width. Note: if this were a right cone, the area of the base would be πr^2.

Next, substitute in for what we know and solve for length:

$$70 = \frac{1}{3}(l)(5)(7) \Rightarrow 6 = l$$

9.2. Formula for the Surface Area of a Sphere

$$A = 4\pi r^2$$

where r stands for the radius of the sphere.

Example 9.2.a: The average radius of the earth is roughly 6371km. Based on this number, what is the surface area of the northern hemisphere assuming that the Earth is a perfect sphere (which it is not)?

A. $8.12x10^7 \pi \, km^2$
B. $4.06x10^7 \pi \, km^2$
C. $1.62x10^8 \pi \, km^2$
D. $3.14x10^7 \pi \, km^2$

Solutions to Example Problems:

Solution to 9.2.a:

The answer is A. Simply plug the radius of the Earth into the formula and cut area in half because it asks for a hemisphere (half the earth):

$$A = \frac{(4)(\pi)(6371)^2}{2} \Rightarrow A = 8.12x10^7\ km^2$$

9.3. Formula for the Volume of a Sphere

$$V = \frac{4}{3}\pi r^3$$

where *r* stands for the radius of the sphere.

Example 9.3.a: Find the radius of a sphere with a volume of 100 cubic units.

A. $75\, units$

B. $\sqrt[3]{\frac{300}{4\pi}} units$

C. $\sqrt[3]{\frac{400}{3\pi}} units$

D. $\frac{75}{\pi} units$

Solutions to Example Problems:

Solution to 9.3.a:

The answer is B. Input what you know and solve for *r*:

$$100 = \frac{4}{3}\pi r^3 \; or \; \frac{300}{4\pi} = r^3 or \sqrt[3]{\frac{300}{4\pi}} = r$$

9.4. Formula for Lateral Surface Area of a Right Circular Cone

---

$$A = \pi r\sqrt{r^2 + h^2}$$

where *r* is the radius of the cone and *h* is the height of the cone. The lateral surface area is all the surface area not including the surface area of the base.

Example 9.4.a: Find the lateral surface area of a right circular cone with a base-radius of 10 cm and a distance from the perimeter of the base to the apex of the cone, s, of 16 cm as shown below.

A. 160π
B. 260π
C. 380π
D. Not enough information given.

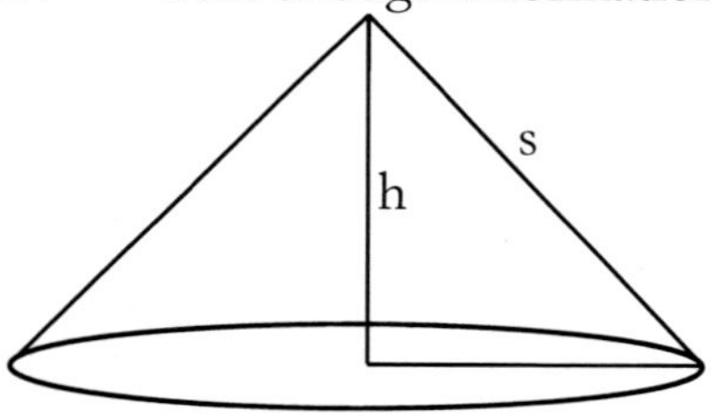

Solutions to Example Problems:

Solution to 9.4.a:

The answer is A. If you dissected the formula above, you may have noticed that $\sqrt{r^2+h^2}$ is exactly the formula for the length of the hypotenuse, *s*, where *h* and *r* are the side lengths (using the Pythagorean theorem). Thus, the lateral surface area of the right circular cone is:

$$A=\pi rs \Rightarrow A=\pi(10)(16) \Rightarrow A=160\pi$$

9.5. Formula for Sum of an Arithmetic Series

$$S_n=\frac{n}{2}\left[2a+(n-1)d\right]=n\left(\frac{a+a_n}{2}\right)$$

where a is the first term in an arithmetic series, and a_n is the nth term in the series.

The variable *d* represents the common difference between each successive term in the series (the fact that there is a common difference between each successive term in the series makes the series arithmetic). For example, if $n_1=1,\ n_2=4,\ and\ n_3=7$, then d=3 for this series.

Example 9.5.a: Find the value of the summation: $\sum_{n=6}^{10} 3n+1$

A. 175
B. 50
C. 100
D. 125

Solutions to Example Problems:

Solution to 9.5.a:

The answer is D. We essentially need the summation between 6 and 10. Therefore, let's find the summation of the first 10 and subtract the first 5 from it (leaving the summation between n=6 and n=10). Using the formula above, we know that we need the first term in the series: a (usually denoted a_1), the 5th term in the series, and the 10th term in the series. To find a_1, $a_5, and\ a_{10}$ we plug in 1, 5, and 10 for *n*:

$$a_1 = 3(1) + 1 = 4$$

$$a_5 = 3(5) + 1 = 16$$

$$a_{10} = 3(10) + 1 = 31$$

Plugging a, a_n, and *n* into the equation, we get our values for s_5 and s_{10} (our 5th and 10th partial sums, respectively):

$$s_5 = n\left(\frac{a + a_n}{2}\right) or\ s_5 = 5\left(\frac{4+16}{2}\right) = 50$$

$$s_{10} = n\left(\frac{a + a_n}{2}\right) or\ s_5 = 10\left(\frac{4+31}{2}\right) = 175$$

Since 175 is the summation of the first 10 terms in the series and 50 is the sum of the first 5 terms in the series, we can subtract the two and get $175 - 50 = 125$.

Alternatively, you could just plug in the numbers 6 through 10 into the summation and add them:

19 + 22 + 25 + 28 + 31 = 125

9.6. Formula for the Sum of a Finite Geometric Series

$$S_n = \frac{a(1 - r^n)}{1 - r}$$

where the a represents the first term in the series, *r* represents the common ratio and *n* is the number of terms in the series.

Finding the common ratio, *r*: if my series of numbers is 1, 3, 9, 27, … then *r*=3 because each term can be multiplied by 3 to get to the next term. Note that a geometric series is a geometric series BECAUSE a common ratio exists between the terms. If no common ratio exists (e.g. 1, 5, 3, 9, 27, 28, 29), then the series is not geometric.

Example 9.6.a: Evaluate: $\sum_{i=1}^{5} 4(-3)^i$

A. -516
B. -732
C. -824
D. -1024

Solutions to Example Problems:

Solution to 9.6.a:

The answer is B. First, we find the first term in the series:

$$a_1 = 4(-3)^1 = -12$$

Next, we find the common ratio by examining the first few terms. The first few terms are:

$$a_1 = -12$$

$$a_2 = 4(-3)^2 = 36$$

$$a_3 = 4(-3)^3 = -108$$

You can get from -12 to 36 and from 36 to -108 by multiplying by -3, therefore, r=-3.

Plugging into our equation, we have:

$$S_n = \frac{a(1-r^n)}{1-r} = \frac{-12(1-(-3)^5)}{1-(-3)} = \frac{-12(1+243)}{4} = (-3)(244) = -732$$

Alternatively, you could just add up all the terms of the series:

-12 + 36 – 108 + 324 – 972 = -732

9.7. Formula for the Sum of an Infinite Geometric Series

$$\sum_{n=0}^{\infty} ar^n = \frac{a}{1-r}, \ |r| < 1$$

where a is the first term in the series, and r is the common ratio. Note that this is an infinite series but even though this series has infinite terms, a finite summation still exists (think of the series $(1, \frac{1}{2}, \frac{1}{3}, \frac{1}{4}, \ldots \frac{1}{n})$ and how each successive term gets smaller and smaller.

Example 9.7.a: Evaluate: $\sum_{n=0}^{\infty} 2\left(\frac{1}{2}\right)^n$

A. ½
B. 2
C. 4
D. 16

Solutions to Example Problems:

Solution to 9.7.a:

The answer is C. First, we find the first term:

$$\sum_{n=0}^{\infty} 2\left(\frac{1}{2}\right)^n = 2\left(\frac{1}{2}\right)^0 = 2$$

Next, we plug this and $r = 1/2$ into the formula:

$$\frac{a}{1-r} = \frac{2}{1-\left(\frac{1}{2}\right)} = 4$$

9.8. Distance Formula

$$d = \sqrt{(x_2 - x_1)^2 + (y_2 - y_1)^2}$$

Please refer to sections **Error! Reference source not found.**or 6.5.

9.9. Midpoint Formula

$$\left(\frac{x_1 + x_2}{2}, \frac{y_1 + y_2}{2}\right)$$

Please refer to section 6.12.

9.10. Formula for Slope

$$m = \frac{dx}{dy} = \frac{rise}{run} = \frac{y_2 - y_1}{x_2 - x_1}$$

Please refer to section **Error! Reference source not found.**.

9.11. Law of Sines

$$\frac{a}{sin A} = \frac{b}{sin B} = \frac{c}{\sin C}$$

where:

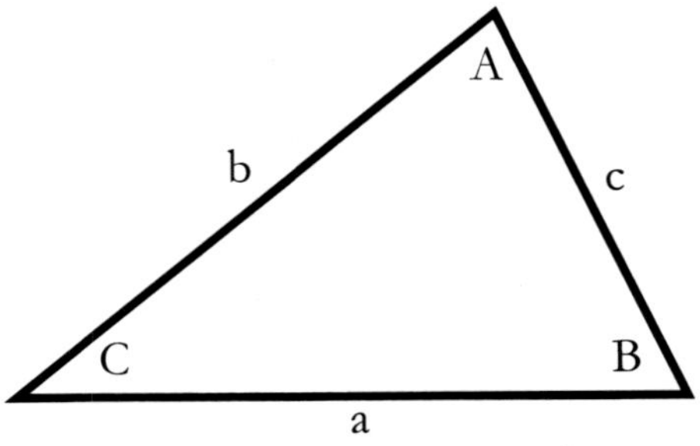

Example 9.11.a: Find the measure of angle B, given that: $a = 5, b = 4, and < A = 35°$.

A. 27.3
B. 28.1
C. 28.9
D. 29.5

Solutions to Example Problems:

Solution to 9.11.a:

The answer is A. Use the formula to set up:

$$\frac{5}{\sin(35)} = \frac{4}{\sin B} \Rightarrow \frac{5}{.574} = \frac{4}{\sin B} \Rightarrow 5(\sin B) = 2.29$$

$$\Rightarrow \sin^{-1}\left(\frac{2.29}{5}\right) = 27.3°$$

9.12. Law of Cosines

$$c^2 = a^2 + b^2 - 2ab\cos C$$

The law of cosines is used for finding a side length when you know the lengths of the other two sides of a triangle and you know the angle opposite your unknown angle.

Example 9.12.a: Find c, given that the information below:

A. 4.36
B. 4.44
C. 4.67
D. 4.75

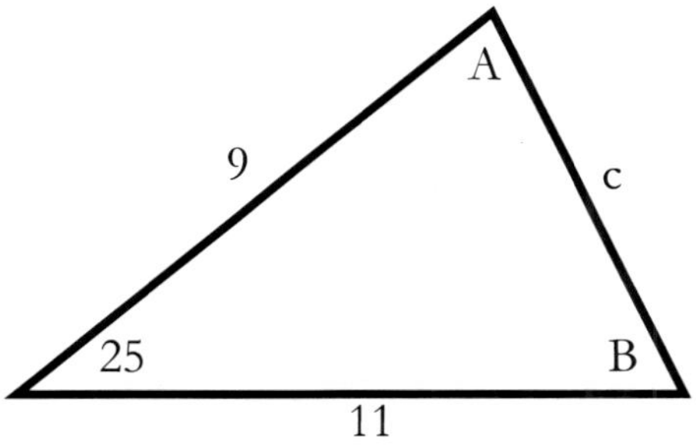

Solutions to Example Problems:

Solution to 9.12.a:

The answer is D. Plug what you know into the formula:

$$c^2 = (11)^2 + (9)^2 - 2(11)(9)\cos 25$$

$$c^2 = 22.55$$

$$c = 4.75$$

9.13. Formula for Variance

$$s^2 = \frac{\sum_{i=1}^{n} (x_1 - \bar{x})^2}{n-1}$$

Please refer to section 7.10.

9.14. Formula for Arc Length

$$s = r\theta$$

Where s represents the arc length, r is the radius of the circle and theta is the angle measurement in radians. Note: multiply radian measurements by $\frac{180°}{\pi}$ to convert them to degree measurements.

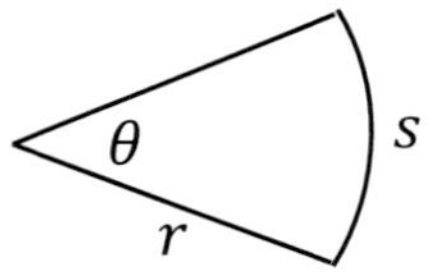

Example 9.14.a: Doni travels 5 km on his unicycle. Given that Doni's unicycle wheel has a radius of 20 cm, how many rotations did his wheel make if he went perfectly straight for the full 5 km?

A. 3627.4 rotations
B. 3978.9 rotations
C. 3894.5 rotations
D. 3988.1 rotations

Solutions to Example Problems:

Solution to 9.14.a:

The answer is B. Note that 2π radians is equal to 1 rotation and note that the arc length is 5 km. To do this, we're going to use stoichiometry to keep track of units because we have to convert several times. First, we find s (which is the arc length) in centimeters.

$$s = 5\,km\,x\frac{1000\,m}{1\,km}x\frac{100\,cm}{1m} = 500,000\,cm$$

Plugging s=500,000cm into the formula, we get:

$$500,000\,cm = 20\,cm(\theta)$$

$$\theta = 25,000\,radians$$

Next we convert from radians to rotations:

$$25,000\,radians\,x\frac{1\,rotation}{2\pi radians} = 3978.9\,rotations$$

9.15. Quadratic Formula

$$\frac{-b \pm \sqrt{b^2 - 4ac}}{2a}$$

Please refer to section 5.11.

Chapter 10

Timed Practice Exam

Please allow yourself exactly 24 minutes to complete the following exam. Answers are provided in the following chapter.

Example 10.1.a: Which is equivalent to $.000000764$?

A. $7.64x10^{-7}$

B. $7.64x10^{7}$

C. $10x7.64^{-7}$

D. $10x\ .000000764$

Example 10.1.b: Luke is making Sloppy Joes. He knows that the correct ratio of ketchup to mustard in the Sloppy Joe sauce is 6:1 ketchup: mustard. How much ketchup should he use to make 8 oz. of Sloppy Joe sauce?

A. $\frac{48}{7}$ oz.

B. 10 oz.

C. $\frac{66}{7}$ oz.

D. $\frac{71}{6}$ oz.

Example 10.1.c: What is the period of the function $f\left(x\right)=3\cos 2\left(\frac{1}{2}x+4\right)+5$?

A. $\frac{1}{2}\pi$

B. 4π

C. π

D. 2π

Example 10.1.d: Which point, if added to the graph of the function, removes the **discontinuity** from $f\left(x\right)=\frac{\sin(x)}{x}$?

A. (0, 1)

B. (0, 0)

C. (1, 0)

D. no point is required, f(x) is already **continuous**

Example 10.1.e: What is the area of the shaded region below given that $f\left(x\right)=-\sqrt{2-x^2}$?

A. 4π

B. 2π

C. π

D. $\pi\sqrt{2}$

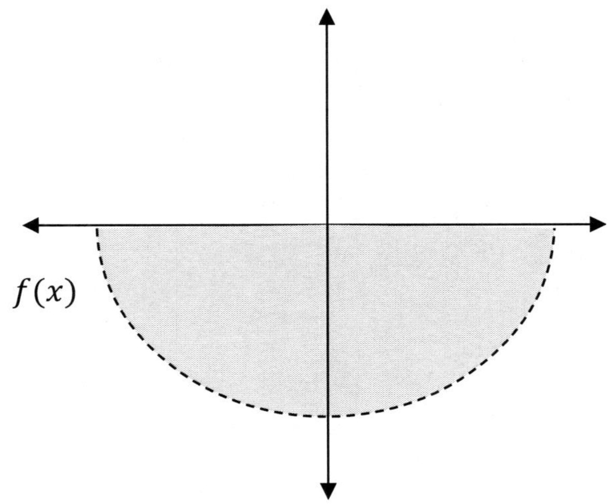

Example 10.1.f: Which of the following represents the number line of the equation:

$$2 > \frac{-2x+4}{3} \geq 6$$

A.

B.

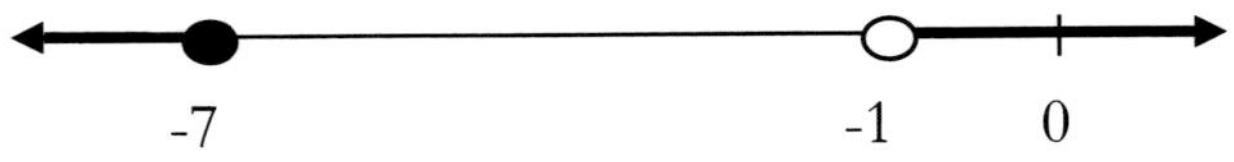

C.

D.

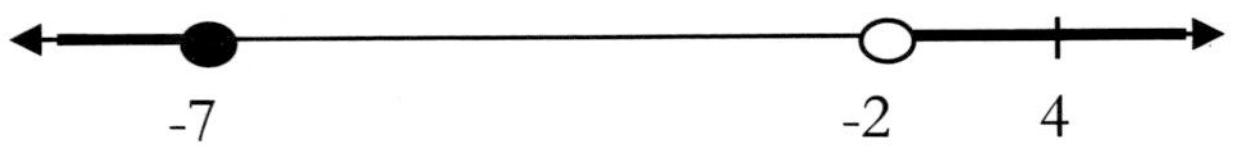

Example 10.1.g: A circular piece of paper with a radius of 5cm is cut from the perimeter to the center of the circle and then overlapped so that an open-topped cone is formed (as shown below where s represents the overlap). Which formula represents the volume of the open-topped cone in terms of s?

A. $V = \frac{1}{3}\left[\pi\left(5 - \frac{s}{2\pi}\right)^2\right]\sqrt{\frac{5s}{\pi} - \frac{s^2}{2\pi^2}}$

B. $V = \frac{1}{3}\left[\pi\left(5 + \frac{s}{2\pi}\right)^2\right]\sqrt{\frac{s}{2\pi} - 5}$

C. $V = \frac{1}{3}\left[\pi\left(5 - \frac{s}{2\pi}\right)^2\right]\sqrt{\frac{5s}{\pi} - \frac{s^2}{4\pi^2}}$

D. $V = \frac{1}{3}\left[\pi\left(5 + \frac{s}{2\pi}\right)^2\right]\sqrt{\frac{5s}{2\pi} + \frac{s^2}{4\pi^2}}$

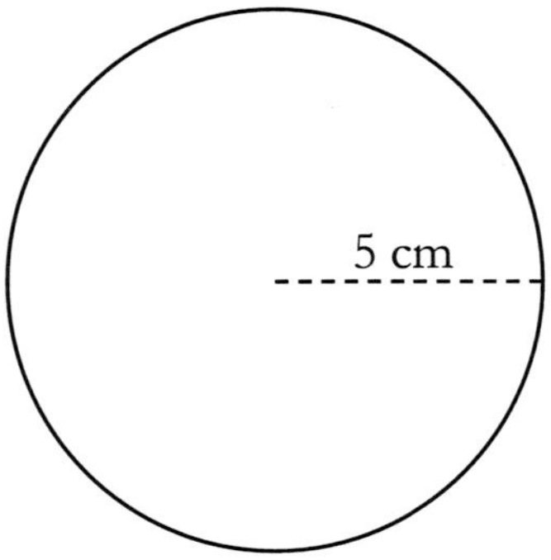

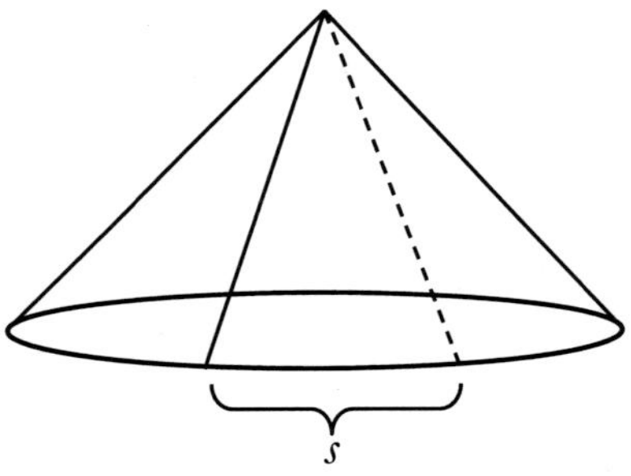

Example 10.1.h: The matrix $\begin{bmatrix} -5 & 4 & 0 \\ 0 & 0 & -6 \end{bmatrix}$ represents triangle ABC. Which matrix represents the triangle after multiplying the matrix by $\begin{bmatrix} 1/2 & 0 \\ 0 & 1/2 \end{bmatrix}$?

A. $\begin{bmatrix} -5 & 4 & 1/2 \\ 1/2 & 1/2 & -6 \end{bmatrix}$

B. $\begin{bmatrix} -5/2 & 2 & 0 \\ 0 & 0 & -3 \end{bmatrix}$

C. $\begin{bmatrix} -5/2 & 4 & 0 \\ 0 & 0 & -3 \end{bmatrix}$

D. Undefined.

Example 10.1.i: Evaluate:

$$\lim_{x \to \infty} \frac{23x^4}{52x^5} + \frac{2}{3} + \frac{3x^2+1}{x^2+2x+1}$$

A. ∞
B. $-\infty$
C. 2/3
D. 11/3

Example 10.1.j: Which is the standard form of the equation: $\frac{(3+i)^2(-i)}{4+i}$

A. $\frac{(-8i+6)}{4+i}$

B. $\frac{3}{14} + \frac{11i}{14}$

C. $\frac{16}{17} + \frac{-38i}{17}$

D. $\frac{1}{13} + \frac{11i}{17}$

Solutions to Example Problems:

Solution to 10.1.a:

The answer is A. Every place a decimal point moves right gives us a 10^{-1} (dividing by 10) while every place that a decimal point moves to the right is a multiplication by 10.

Solution to 10.1.b:

The answer is A. First you set up the following ratios just to make sure you have it straight:

$$\frac{ketchup}{mustard}=\frac{6}{1}\ and\ \frac{ketchup}{total\ sauce}=\frac{6}{7}$$

Next, substitute the total number of ounces you need into the total sauce part of the ratio:

$$\frac{ketchup}{8}=\frac{6}{7}$$

Cross multiply, and solve for ketchup:

$$7(ketchup)=48 \Rightarrow ketchup=\frac{48}{7}$$

Solution to 10.1.c:

The answer is D. First, we must manipulate the function to represent the standard form equation:

$$f(x)=\mathrm{A}\ \sin B(x-C)+D$$

where A is the amplitude, $\frac{2\pi}{B}$ is the period, C is the horizontal shift (a negative C indicates a shift to the right and a positive C indicates a shift to the left) and D is the vertical shift.

To do that, we have to factor $\frac{1}{2}$ out of the inside of the parentheses:

$$f(x)=3\cos 2\left(\frac{1}{2}\right)(1x+8)+5 \Rightarrow f(x)=3\cos 1(x+8)+5$$

B=1, which we plug into the formula for period: $\frac{2\pi}{B}=period$. Thus, the period is 2π.

Solution to 10.1.d:

The answer is A. The only point where $f(x)=\frac{\sin(x)}{x}$ is not continuous is where x equals 0. If you graph the function using a graphing calculator in radians mode, you can tell that the point (0, 1) is the only point that does not exist. Thus, if you add the point (0, 1) to the graph of the function, it becomes continuous again.

Solution to 10.1.e:

The answer is C. This is a semi-circle, which can be seen by taking the equation: $y=-\sqrt{2-x^2}$; squaring both sides: $y^2=2-x^2$; and moving the terms around: $y^2+x^2=2$. Hopefully you can see that this is a circle centered at (0,0) with a radius of $\sqrt{2}$ (the general from equation for a circle is $x^2+y^2=r^2$). The area of the entire circle would be $A=\pi r^2$ or $A=\pi\left(\sqrt{2}\right)^2$ or $A=2\pi$. However, since we only have half that circle, the real area is π.

Solution to 10.1.f:

The answer is B. With a double inequality, what you do to the middle in order to solve for x must be done to each side of the **inequality**.

First, multiply by 3: $6>-2x+4\geq 18$

Then, subtract 4: $2>-2x\geq 14$

Last, divide by -2 and remember to switch the direction of the inequalities: $-1<x\leq -7$. x is greater than -1 and less than or equal to -7.

Solution to 10.1.g:

The answer is C. First, look at the equation for the volume of a cone: $V=\frac{1}{3}Bh$ where *B* is the base of the cone and *h* is the height. This question is tricky because both the base of the cone and the height of the cone change as the overlap, *s*, changes; we must come up with equations for height and base, each in terms of *s*.

First, let's find a formula for the base of the cone in terms of *s*. The base of a cone is a circle with area: $A=\pi r^2$. However, the radius of our circle will change as *s* changes.

The perimeter of the original circle is $2\pi r=10\pi$. Therefore, the perimeter of the base of the cone is $10\pi-s$. We can use this to find the radius of the base of the cone in terms of *s*:

$$2\pi r=10\pi-s$$

$$r=\frac{10\pi-s}{2\pi}\Rightarrow r=5-\frac{s}{2\pi}$$

Substituting *r* from the line above into the formula for the base of the cone ($A=\pi r^2$) we get:

$$A=\pi\left(5-\frac{s}{2\pi}\right)^2$$

Next, let's come up with an equation which represents the height of the cone. If we lay the cone directly on its side, we get an isosceles triangle with the following dimensions:

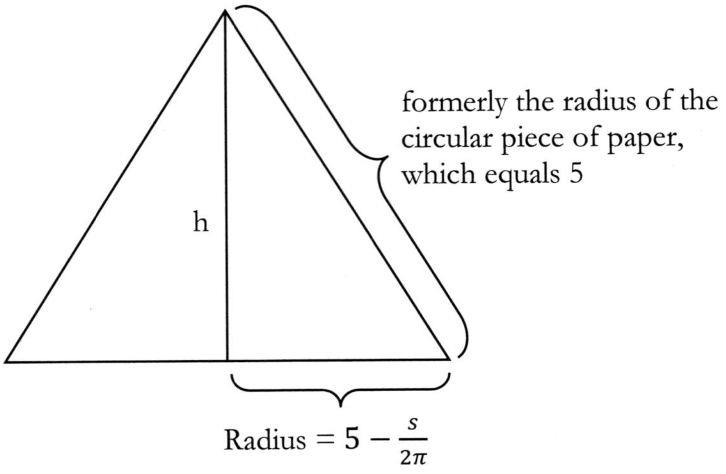

Next, we solve for *h* using the Pythagorean theorem where $5-\dfrac{s}{2\pi}$ is one leg and 5 is the hypotenuse:

$$\left(5-\frac{s}{2\pi}\right)^2 + h^2 = 5^2$$

$$\Rightarrow 25-\frac{5s}{\pi}+\frac{s^2}{4\pi^2}+h^2 = 25$$

$$\frac{5s}{\pi}-\frac{s^2}{4\pi^2} = h^2$$

$$\Rightarrow h = \sqrt{\frac{5s}{\pi}-\frac{s^2}{4\pi^2}}$$

Since we finally have the height in terms of *s* and the area of the base in terms for *s*, we can plug them into the equation for the volume of a right cone, $V = \dfrac{1}{3}Bh$:

$$V = \frac{1}{3}\left[\pi\left(5-\frac{s}{2\pi}\right)^2\right]\sqrt{\frac{5s}{\pi}-\frac{s^2}{4\pi^2}}$$

Solution to 10.1.h:

The answer is D. You can only multiply matrices when the number of columns of the first matrix matches the number of rows in the second matrix. An easy way to test for this is to write the matrices out as such (I will use the matrices from this example.): the first matrix is a 2x3 and the second matrix is a 2x2. When you line them up: 2x3 and 2x2, the middle two numbers must be equal (in this case, 3 and 2 are not equal). Thus you cannot multiply them out.

Note, if you can multiply two matrices like a 4x5 and a 5x2, the size of the resulting matrix will be the two outside numbers in order. In this case, the resulting matrix would be a 4x2 matrix. The order in which matrices are multiplied is vitally important.

Solution to 10.1.i:

The answer is D. The limit of the first term approaches 0 because $52x^5$ (the denominator) increases in size much faster than its numerator $23x^4$. The second term stays at 2/3. In regards to the third term, when a limit is taken to infinity, only the highest degree terms on the top and the bottom are considered (every other term becomes trivial at very high numbers); thus, the fraction reduces to $\frac{3x^2}{x^2}$ or simply 3. Thus we have 0 + 2/3 + 3 which is 11/3.

Solution to 10.1.j:

The answer is C. First, we simplify the top. Remember that $i^2 = -1$

$$\frac{(3+i)^2\,(-i)}{4+i} = \frac{(3+i)(3+i)(-i)}{4+i} = \frac{(9+6i+i^2)(-i)}{4+i} = \frac{(9+6i-1)(-i)}{4+i} = \frac{(8+6i)(-i)}{4+i} =$$

$$\frac{(-8i-6i^2)}{4+i} = \frac{(-8i+6)}{4+i}$$

Next, we multiply the top and the bottom of the fraction by the conjugate of $4 + i$ to get i out of the denominator.

$$\frac{(6-8i)}{4+i} * \frac{(4-i)}{(4-i)} = \frac{(24-38i+8i^2)}{(16-i^2)} = \frac{(24-38i-8)}{(16+1)} = \frac{(16-38i)}{17} = \frac{16}{17} - \frac{38i}{17}.$$

Index

Notes:

Made in United States
Cleveland, OH
05 November 2026

47763602R00108